RAND NATIONAL DEFENSE RESEARCH INSTITUTE

Operating Under a Continuing Resolution

A Limited Assessment of Effects on Defense Procurement Contract Awards

Stephanie Young and J. Michael Gilmore

Prepared for the Office of the Secretary of Defense

For more information on this publication, visit www.rand.org/t/RR2263

Library of Congress Cataloging-in-Publication Data is available for this publication.
ISBN: 978-1-9774-0103-8

Published by the RAND Corporation, Santa Monica, Calif.

Preface

Under the normal legislative process, Congress completes 12 regular appropriation bills each fiscal year to fund the activities of federal agencies. When Congress is unable to pass these appropriation bills before the beginning of a fiscal year, a continuing appropriation act (or continuing resolution [CR]) can be enacted as a joint resolution to provide stopgap funding at a specified rate and for a specified period of time. In recent years, operating under a CR at the start of a fiscal year has become the norm for federal agencies. There is general agreement among those who have studied them that CRs, because they limit government agencies' ability to engage in activities for which authority has not been granted by law, cause inefficiency in government operations and therefore adds costs.

This research report provides an overview of issues associated with U.S. Department of Defense (DoD) operations under a CR and performs a limited assessment of whether particular consequences that DoD and its leadership cite as occurring broadly as a result of operating under CRs have actually been occurring. In particular, the report explores whether there is evidence that delays in making procurement awards for weapons and cost increases resulting from those delays are occurring broadly and are associated with CRs because of CRs' constraints on funding and on initiating new activities. The results of our analysis are mixed. They do not provide strong evidence that CRs are causing delays and cost increases; because of their limitations, however, our results also do not provide definitive evidence that such negative effects are *not* occurring. The report documents preliminary research and should be considered only a first, limited step in addressing the important and broad policy issue of the consequences of operating under a CR rather than under a regular appropriation.

This work was conducted between about March and December 2017 and should be of interest to senior decisionmakers and acquisition professionals across DoD. Other recent RAND work related to analysis of drivers of acquisition performance outcomes includes *Program Characteristics That Contribute to Cost Growth*, *Quantifying Cost and Schedule Uncertainty for Major Defense Acquisition Programs (MDAPs)*,

Extreme Cost Growth, Developing a Methodology for Risk-Informed Trade-Space Analysis in Acquisition.[1]

This research was sponsored by the Office of the Under Secretary of Defense for Acquisition, Technology and Logistics Acquisition Resources and Analysis Directorate and conducted within the Acquisition and Technology Policy Center of the RAND National Defense Research Institute, a federally funded research and development center sponsored by the Office of the Secretary of Defense, the Joint Staff, the Unified Combatant Commands, the Navy, the Marine Corps, the defense agencies, and the defense Intelligence Community.

For more information on the RAND Acquisition and Technology Policy Center, see www.rand.org/nsrd/ndri/centers/atp or contact the director (contact information is provided on the webpage).

---

[1] Mark A. Lorell, Leslie Adrienne Payne, and Karishma R. Mehta, *Program Characteristics That Contribute to Cost Growth: A Comparison of Air Force Major Defense Acquisition Programs*, Santa Monica, Calif.: RAND Corporation, RR-1761-AF, 2017; Thomas Light, Robert S. Leonard, Julia Pollak, Meagan L. Smith, and Akilah Wallace, *Quantifying Cost and Schedule Uncertainty for Major Defense Acquisition Programs (MDAPs)*, Santa Monica, Calif.: RAND Corporation, RR-1723-AF, 2017; Mark A. Lorell, Robert S. Leonard, and Abby Doll, *Extreme Cost Growth: Themes from Six U.S. Air Force Major Defense Acquisition Programs*, Santa Monica, Calif.: RAND Corporation, RR-630-AF, 2015; Craig A. Bond, Lauren A. Mayer, Michael McMahon, James G. Kallimani, and Ricardo Sanchez, *Developing a Methodology for Risk-Informed Trade-Space Analysis in Acquisition*, Santa Monica, Calif.: RAND Corporation, RR-701-A, 2015.

Contents

Figure and Tables

Figure

Tables

Summary

Background

Under the normal legislative process, Congress would complete 12 regular appropriation bills each fiscal year (FY) to fund the activities of most federal agencies. When Congress is unable to pass appropriation bills before the beginning of an FY, a continuing appropriation act (or continuing resolution [CR]) can be enacted as a joint resolution to provide stopgap funding at a specified rate and for a specified period of time. In recent years, federal agencies have often operated under a CR at the beginning of an FY, sometimes for several months.

There is general agreement among those who have studied them that CRs cause inefficiency in government operations and have other harmful effects. There is also agreement that these harmful effects could be associated with costs for government personnel's work that would not otherwise be done, for delays in changing the course of programs or eliminating outmoded programs, and costs for delayed procurements, among other factors. However, as the U.S. Government Accountability Office has noted, data and other information are lacking that would permit analysts to be able to derive defensible estimates of those costs and to draw general conclusions.[1]

Approach

This report focuses on only two of the many potential deleterious effects of CRs noted in past, generally testimony-based, assessments: delays and increases in the costs of planned procurements—in particular, weapon system procurements made by the U.S. Department of Defense (DoD). We chose this focus for our analysis because data are available to examine it: the budget material displaying projected (and ultimately realized) procurement award dates and unit costs composing successive DoD Presi-

[1] U.S. Government Accountability Office, *Continuing Resolutions: Uncertainty Limited Management Options and Increased Workload in Selected Agencies—Report to the Ranking Member, Subcommittee on Oversight of Government Management, the Federal Workforce, and the District of Columbia, Committee on Homeland Security and Governmental Affairs, U.S. Senate*, Washington, D.C., GAO-09-879, September 2009.

dent's budget submissions. In the past several years, DoD has prepared and submitted requests for specific relief (*anomalies*) from the constraints that CRs impose on new activities. Some of those requests have expressed concern that CRs could cause delays in making weapon procurement awards under new or existing contracts and thereby increase costs. Moreover, DoD leaders have indicated on several occasions these constraints and other deleterious effects of CRs are widespread.

Results

In our limited analysis, we were not able to identify a strong connection between CRs and delays and cost increases in procurement awards. However, the limitations in our analysis mean that it also does not provide definitive evidence that such negative effects are *not* occurring. In particular, because our data sets are small, our analysis enables the statistical significance of effects in some cases to be determined only when effects are relatively large. With those caveats, the highlights of our results include the following:

1. Between FY 2013 and FY 2015, irrespective of whether a CR occurred when an award was planned to be made, unit-cost decreases were more likely than increases. Overall, delays were experienced about as often as no change was experienced in award dates or accelerations.
2. Between FY 2013 and FY 2015, realized unit costs relative to original projections declined somewhat on average for all awards examined. There was no statistically significant difference between the cost declines experienced for procurement awards planned originally during periods subsequently affected by CRs and those for awards planned during periods not affected by CRs.
3. Between FY 2013 and FY 2015, the occurrence and length of a delay in making an award did not depend on whether an award was planned to occur during a CR period or a non-CR period.
4. When they occurred, delays in awards were significantly longer for awards planned originally for FYs 2013–2015 than during FY 1999. Nonetheless, there is no significant difference in proportion of programs experiencing delays, or overall net changes in award dates, between awards planned for FYs 2013–2015 (which were affected by longer CRs) and awards planned to be made during FY 1999 (which operated under CRs for only three weeks). Cost increases were, on average, significantly greater for awards planned during FY 1999 than for awards planned for FYs 2013–2015.

By comparing FY 1999 awards, when a very short period of CRs occurred, with FY 2013–2015 awards, we have attempted to control for the possibility that program managers could be adjusting cost estimates and award dates systematically in anticipation that CRs will occur. However, many factors for which we cannot control affect the timing and content of DoD's contract awards, and they could be masking the negative effects of CRs. These factors include delays (from initially expected schedules) in completing DoD's multiple internal review processes, flexibilities that might be available and used under certain circumstances for awards under existing contracts to be made during CRs, and flexibilities for other awards under new contracts to be made for programs already in production and procurement. Our data set also did not include a substantial number of procurement awards planned under new contracts and, in particular, because it focused on procurement contracting, excluded awards for research, development, test, and evaluation efforts, including new-start research, development, test, and evaluation contracts. Therefore, the lack of conclusive evidence found in our analysis does not support a definitive conclusion that the widely expressed concerns regarding CRs' effects are invalid.

Acknowledgments

We acknowledge the support of the sponsor for this work within the Office of the Under Secretary of Defense for Acquisition, Technology and Logistics, Nancy Spruill, as well as the support of Cynthia R. Cook, the former director of the Acquisition and Technology Policy Center within the RAND National Defense Research Institute. We also thank Christopher A. Mouton, Irv Blickstein, Philip S. Anton, and Claude Messan Setodji of RAND, as well as our external reviewer, Philip J. Candreva of the Naval Postgraduate School, for providing very helpful comments.

Abbreviations

AAE	Army Acquisition Executive
APB	acquisition program baseline
CR	continuing resolution
CRS	Congressional Research Service
DAE	Defense Acquisition Executive
DoD	U.S. Department of Defense
ERI	European Reassurance Initiative
FRP	full-rate production
FY	fiscal year
GAO	U.S. Government Accountability Office
GRRIP	Global Rapid Response Information Package
JASSM	Joint Air-to-Surface Standoff Missile
JSF	Joint Strike Fighter
LCS	littoral combat ship
MYP	multiyear procurement
O&M	operation and maintenance
OMB	Office of Management and Budget
RAM	Rolling Airframe Missile
RDT&E	research, development, test, and evaluation
SAR	Selected Acquisition Report
WSARA	Weapon Systems Acquisition Reform Act of 2009

Operating Under a Continuing Resolution: An Assessment of Effects on Defense Procurement Contract Awards

This report provides an overview of issues associated with U.S. Department of Defense (DoD) operations under continuing resolutions (CRs) and performs an assessment focusing on particular consequences that DoD cites of operating under a CR. The particular consequences the report explores are whether CRs are associated with delays in making procurement contract awards for weapons that lead to increased costs due to a CR's constraints on initiating new activities. The report documents limited-scope preliminary research and should be considered only a first step in addressing the important and broad policy issue of the effects that operating under a CR (rather than under a regular appropriation) have on the government. More research should be done to better understand CRs' effects on acquisition programs.

Overview of Continuing Resolutions

Under the normal legislative process, Congress completes 12 regular appropriation bills each fiscal year (FY) to fund the activities of most federal agencies. When Congress is unable to pass appropriation bills before the beginning of an FY, a CR (also known as a *continuing appropriation act*) can be enacted as a joint resolution to provide stopgap funding at a specified rate and for a specified period of time. Each of these acts is formally titled, "An appropriation act that provides budget authority for federal agencies, specific activities, or both to continue in operation when Congress and the President have not completed action on the regular appropriation acts by the beginning of the fiscal year."[1] The legislation indicates the CR's coverage, the specific projects and activities for which it provides appropriations.[2]

[1] U.S. Government Accountability Office (GAO), *Principles of Federal Appropriations Law*, 3rd ed., Vol. II, Washington, D.C., GAO-06-382SP, February 1, 2006, p. 8-2. These acts can be part of other legislation with other names (see reference list for examples).

[2] James V. Saturno and Jessica Tollestrup, *Continuing Resolutions: Overview of Components and Recent Practices*, Washington, D.C.: Congressional Research Service, R42647, January 14, 2016, pp. 3–4.

A CR can be implemented as either an interim measure (as short as a day) or a measure providing funding for a full FY. An interim CR is a stopgap measure before passage of a regular appropriation bill or, if no regular appropriation bill is anticipated, as a measure to provide appropriations for the remainder of the FY. A CR's duration is the period of time for which the legislation provides appropriations. In practice, the period ends either at the time indicated in the CR or upon enactment of a regular appropriation—whichever comes first.[3] The length of a CR has significant implications for the ease of managing an acquisition program while operating under a CR. As we emphasize later in this report, by many accounts, managers have developed approaches for mitigating risks associated with CRs for a matter of weeks or even months, but the longer the interim solution remains in place, the fewer the options remain available.

Table 1.1 describes the characteristics of CRs since 2011, including the legislation enacting each CR, the enactment date and expiration date indicated in the legislation, and the duration, in days. (Figure 1.1 provides a graphical depiction of the duration of these CRs over time.) Each of these FYs begins with enactment of a CR, a practice that has become the norm. The longest overall period of time operating under a CR in these years was 2011, which experienced seven relatively short CRs for a total of almost seven months. Before 2017, the longest single CRs occurred in 2013 (six months) and 2014 (four months). On average, since 2011, DoD operated under a CR for an average of five months. In a given FY, about 3.7 CRs passed, on average, and their average duration was about one month (37 days).

Unlike a regular appropriation act, a CR generally does not provide specific amounts of budget authority. Rather, a CR generally provides "such amounts as may be necessary" at a given specified funding rate (often established in reference to a "current rate" established at the prior year's appropriation level).[4] Several common formulas related to the current rate have been used in past CRs, including "at the current rate," "not to exceed the current rate," and whichever is larger between a current rate and the rate provided in another reference item.[5] For example, the first CR for FY 2017 provided budget authority for projects and activities at FY 2016 levels, minus 0.496 percent.[6] The rate under a CR, then, is usually the level established in the previous FY less whatever reduction Congress wishes to impose.

The relatively rigid formulation of a CR and the exigency of maintaining congressional intent in appropriations have been interpreted as prohibiting the use of funds

---

[3] Saturno and Tollestrup, 2016, p. 4.

[4] GAO, 2006, pp. 8-6, 8-10.

[5] GAO, 2006, pp. 8-6, 8-10–8-14, provides a detailed discussion of the interpretation and use of these different rates for operation.

[6] Pub. L. 114-53, TSA Office of Inspection Accountability Act of 2015, September 30, 2015; Lynn M. Williams and Darren P. Wees, *FY2017 Defense Spending Under an Interim Continuing Resolution (CR): In Brief*, Washington, D.C.: Congressional Research Service, R44636, November 7, 2016.

Table 1.1
Recent Continuing Resolutions Affecting the Department of Defense, Fiscal Years 2011–2017

Fiscal Year	Public Law	Title	Enactment Date	Expiration Date	Duration, in Days
2011	111-242	Continuing Appropriations Act, 2011	Sept. 30, 2010	Dec. 3, 2010	64
2011	111-290	Joint Resolution Making Further Continuing Appropriations for Fiscal Year 2011, and for Other Purposes	Dec. 4, 2010	Dec. 18, 2010	15
2011	111-317	Joint Resolution Making Further Continuing Appropriations for Fiscal Year 2011, and for Other Purposes	Dec. 18, 2010	Dec. 21, 2010	3
2011	111-322	Continuing Appropriations and Surface Transportation Extensions Act, 2011	Dec. 22, 2010	Mar. 4, 2011	73
2011	112-4	Furthering Continuing Appropriations Amendments, 2011	Mar. 2, 2011	Mar. 18, 2011	14
2011	112-6	Additional Continuing Appropriations Amendments, 2011	Mar. 18, 2011	Apr. 8, 2011	21
2011	112-8	Further Additional Continuing Appropriations Amendments, 2011	Apr. 9, 2011	Apr. 15, 2011	7
2012	112-33	Continuing Appropriations Act, 2012	Sept. 30, 2011	Oct. 4, 2011	4
2012	112-36	Continuing Appropriations Act, 2012	Oct. 5, 2011	Nov. 18, 2011	45
2012	112-55	Consolidated and Further Continuing Appropriations Act, 2012	Nov. 18, 2011	Dec. 16, 2011	28
2012	112-67	Joint Resolution Making Further Continuing Appropriations for Fiscal Year 2012, and for Other Purposes	Dec. 16, 2011	Dec. 17, 2011	1
2012	112-68	Joint Resolution Making Further Continuing Appropriations for Fiscal Year 2012, and for Other Purposes	Dec. 17, 2011	Dec. 23, 2011	6
2013	112-175	Continuing Appropriations Resolution, 2013	Sept. 28, 2012	Mar. 27, 2013	178
2014	113-39	Pay Our Military Act	Sept. 30, 2013	—[a]	17[a]
2014	113-44	Department of Defense Survivor Benefits Continuing Appropriations Resolution, 2014	Oct. 10, 2013	Dec. 15, 2013[a]	8[a]
2014	113-46	Continuing Appropriations Act, 2014	Oct. 17, 2013	Jan. 15, 2014[a]	107[a]
2014	113-73	Joint Resolution Making Further Continuing Appropriations for Fiscal Year 2014, and for Other Purposes	Jan. 15, 2014	Jan. 18, 2014[a]	3[a]
2015	113-164	Continuing Appropriations Resolution, 2015	Sept. 19, 2014	Dec. 11, 2014	72
2015	113-202	Joint Resolution Making Further Continuing Appropriations for Fiscal Year 2015, and for Other Purposes	Dec. 12, 2014	Dec. 13, 2014	1

Table 1.1—Continued

Fiscal Year	Public Law	Title	Enactment Date	Expiration Date	Duration, in Days
2015	113-203	Joint Resolution Making Further Continuing Appropriations for Fiscal Year 2015, and for Other Purposes	Dec. 13, 2014	Dec. 17, 2014	4
2016	114-53	TSA Office of Inspection Accountability Act of 2015	Sept. 30, 2015	Dec. 11, 2015	72
2016	114-96	Further Continuing Appropriations Act, 2016	Dec. 11, 2015	Dec. 16, 2015	5
2016	114-100	Joint Resolution Making Further Continuing Appropriations for Fiscal Year 2016, and for Other Purposes	Dec. 16, 2015	Dec. 22, 2015	6
2017	114-223	Continuing Appropriations and Military Construction, Veterans Affairs, and Related Agencies Appropriations Act, 2017, and Zika Response and Preparedness Act	Sept. 29, 2016	Dec. 9, 2016	71
2017	114-254	An Act Making Appropriations for Energy and Water Development and Related Agencies for the Fiscal Year Ending September 30, 2016, and for Other Purposes	Dec. 12, 2016	Apr. 28, 2017	137
2017	115-30	Joint Resolution Making Further Continuing Appropriations for Fiscal Year 2017, and for Other Purposes	Apr. 28, 2017	May 5, 2017	7

SOURCES: Saturno and Tollestrup, 2016; Nese F. DeBruyne, *Defense Authorization and Appropriations Bills: FY1970–FY2017*, Washington, D.C.: Congressional Research Service, January 18, 2017; "Appropriations for Fiscal Year 2017," *Congress.gov*, last updated May 4, 2017.

[a] According to Saturno and Tollestrup, 2016, p. CRS-30,

> A total of four CRs were enacted for FY2014. This count includes two CRs that provided funding for only specific programs and activities during the FY2014 funding gap. The Pay Our Military Act (P.L. 113-39) was enacted on September 30, 2013, and provided funding for FY2014. The Department of Defense Survivor Benefits Continuing Appropriations Resolution, 2014 (P.L. 113-44), was enacted on October 10, 2013, and expired on December 15, 2013. However, the funding provided by both of these CRs was terminated on October 17, 2013, through the enactment of a third CR, P.L. 113-46, which broadly funded the previous fiscal year's activities through January 15, 2014. The funding provided by this third CR was extended through January 18 by the enactment of a fourth CR (P.L. 113-73). Section 118 of P.L. 113-46 provided that the time covered by that act was to have begun on October 1, 2013. . . . [The] duration in days for the first two CRs is considered to have ended on October 17, 2013. The third CR is considered to have begun on October 1, 2013, and expired on January 15, 2014.

for "new starts." For example, recent CR language has included a provision prohibiting "any project or activity for which appropriations, funds, or other authority were not available" in the previous FY.[7] The interpretation of this prohibition in specific cases of budget execution has proven to be a matter of legal debate.[8]

---

[7] GAO, 2006, pp. 8-24–8-25.

[8] GAO, 2006, pp. 8-24–8-25.

Figure 1.1
Length of Continuing Resolutions, in Days, Fiscal Years 2011 Through 2017

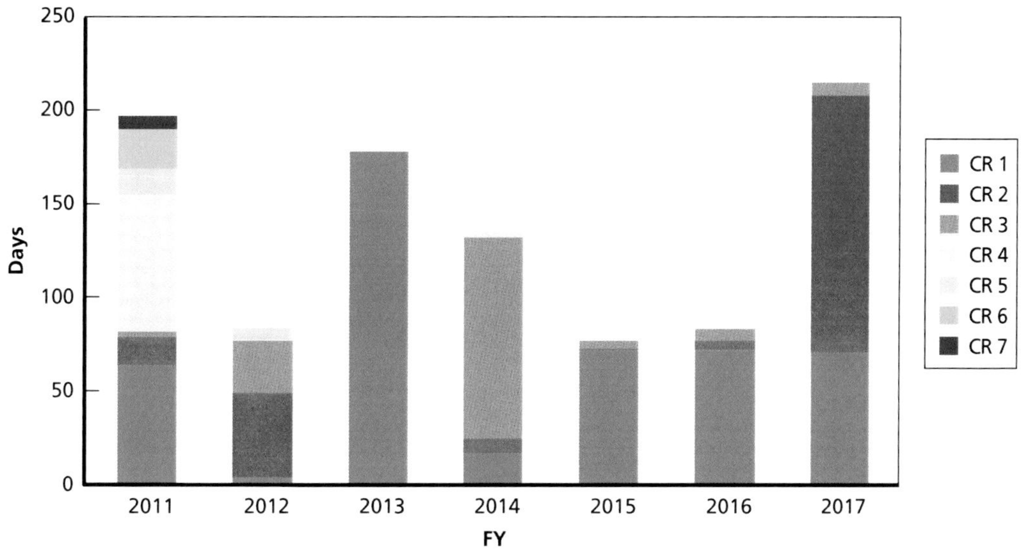

RAND *RR2263-A.1*

One exception to the relative rigidity of a CR is the provision for anomalies. Such provisions delineate exceptions to the CR's duration, rate, or coverage.[9] One observer has called them the "exceptions to a CR's steady-as-it-goes approach to funding the government."[10] Another described them as "special cases that the agencies put forward . . . asking Congress for special dispensation to fund programs."[11] The leadership of a department or agency can submit requests to the Office of Management and Budget (OMB) for anomalies that the department or agency would like to see included in the CR. In recent years, DoD has requested anomalies, for example, to support production increases, multiyear procurement (MYP), and new starts. In most cases, these requests for anomalies do not make it into the legislation. Despite sending dozens of anomalies to OMB, the Pentagon has only rarely succeeded in getting them into legislation.[12] This is especially the case when the CR is anticipated to be of short duration.

Information that DoD provided to RAND regarding anomaly requests prepared within DoD suggests that, with the exception of FY 2017, the department has not been successful in convincing OMB and Congress to include anomalies granting it exceptions from CR-imposed constraints on initiating new activities (see Table 1.2). Pre-

---

[9] Saturno and Tollestrup, 2016, p. 7.

[10] David Hawkings, "What Is a Continuing Resolution?" *Roll Call*, September 7, 2016.

[11] Committee on Armed Services, U.S. House of Representatives, *Damage to the Military from a Continuing Resolution: Hearing Before the Committee on Armed Services*, hearing, Washington, D.C., April 5, 2017.

[12] Darren Samuelsohn, "Welcome to CR Hell," *Politico*, October 21, 2015.

Table 1.2
Department of Defense–Prepared Anomaly Requests Compared with Anomalies Appearing in Continuing Resolutions for Fiscal Years 2013 Through 2017

FY	DoD Anomaly Requests Identified[a]	Requests Appearing in Legislation	Percentage Appearing in Legislation
2013	18	0	0
2014	66[b]	0	0
2015	21	2	10
2016	34	1	3
2017	249[c]	10	4
Totals	388	13	3

[a] Data received from DoD include requests for anomalies prepared by DoD staff for potential submission to OMB but not the final anomalies approved by DoD leadership and submitted to OMB. We do not have access to the anomalies submitted by OMB to Congress.

[b] We did not receive a complete list of requests for FY 2014, only summary numbers of requests and select expiring authority requests.

[c] This total likely overstates the number of requests submitted by DoD to OMB for FY 2017. We received long lists of anomalies prepared by DoD staff that appear to be predecisional (i.e., not approved by DoD leadership for submission to OMB).

liminary analysis suggests that only a small fraction (about 3 percent) of the anomaly requests prepared by DoD staff and provided to RAND have been incorporated in the CRs enacted from FY 2013 to FY 2017.[13]

However, anomalies have been granted to resource certain priorities or to provide funding when appropriators anticipated a long-duration CR. For insight into the kinds of requests granted as anomalies in recent practice, see Table 1.3, which displays legislation between FY 2011 and FY 2017. It shows, for example, that, in the first quarter of FY 2012, an anomaly was granted to support the time-sensitive need to resource the dynamic requirements of redeploying all U.S. troops from Iraq.[14] One of the anomalies included in the first quarter of FY 2012 provided authority to spend existing funds for the Office of Security Cooperation–Iraq but did not provide additional resources. Congress granted DoD this authority in an anomaly just one day before the office (a critical component of U.S. transition planning in Iraq) was scheduled to become

[13] The data source here is annual anomaly requests provided by the Office of the Under Secretary of Defense for Acquisition, Technology and Logistics. Our team had access to limited information regarding anomaly requests. Definitive analysis of anomaly requests and their final disposition would require additional data.

[14] See "Impact of Continuing Resolutions and Funding Delays" in Rick Brennan, Jr., Charles P. Ries, Larry Hanauer, Ben Connable, Terrence Kelly, Michael McNerney, Stephanie Young, Jason H. Campbell, and K. Scott McMahon, *Ending the U.S. War in Iraq: The Final Transition, Operational Maneuver, and Disestablishment of United States Forces–Iraq*, Santa Monica, Calif.: RAND Corporation, RR-232-USFI, 2013, pp. 116–119.

Table 1.3
Anomalies Appearing in Continuing Resolutions Affecting the Department of Defense, Fiscal Years 2011–2016

FY	Public Law	Enactment Date	Expiration Date	DoD Anomaly
2011	111-242	September 30, 2010	December 3, 2010	None
2011	111-290	December 4, 2010	December 18, 2010	None
2011	111-317	December 18, 2010	December 21, 2010	None
2011	111-322	December 22, 2010	March 4, 2011	None
2011	112-4	March 2, 2011	March 18, 2011	None
2011	112-6	March 18, 2011	April 8, 2011	None
2011	112-8	April 9, 2011	April 15, 2011	None
2012	112-33	September 30, 2011	October 4, 2011	"Notwithstanding section 101, amounts made available by this Act for 'Department of Defense—Operation and Maintenance—Operation and Maintenance, Air Force' may be used by the Secretary of Defense for operations and activities of the Office of Security Cooperation in Iraq and security assistance teams, including life support, transportation and personal security, and facilities renovation and construction . . . ." (§ 116) "Notwithstanding section 101, funds made available in title IX of division A of Public Law 112-10 [Department of Defense and Full-Year Continuing Appropriations Act, 2011, April 15, 2011] for 'Overseas Contingency Operations' shall be available at a rate for operations not to exceed the rate permitted by H.R. 2219 (112th Congress) as passed by the House of Representatives on July 8, 2011." (§ 117)
2012	112-36	October 5, 2011	November 18, 2011	"Notwithstanding section 101, amounts made available by this Act for 'Department of Defense—Operation and Maintenance—Operation and Maintenance, Air Force' may be used by the Secretary of Defense for operations and activities of the Office of Security Cooperation in Iraq and security assistance teams, including life support, transportation and personal security, and facilities renovation and construction . . . ." (§ 116) "Notwithstanding section 101, funds made available in title IX of division A of Public Law 112-10 for 'Overseas Contingency Operations' shall be available at a rate for operations not to exceed the rate permitted by H.R. 2219 (112th Congress) as passed by the House of Representatives on July 8, 2011." (§ 117)
2012	112-55	November 18, 2011	December 16, 2011	None

Table 1.3—Continued

FY	Public Law	Enactment Date	Expiration Date	DoD Anomaly
2012	112-67	December 16, 2011	December 17, 2011	None
2012	112-68	December 17, 2011	December 23, 2011	None
2013	112-175	September 28, 2012	March 27, 2013	"No appropriation or funds made available or authority granted pursuant to section 101 for the Department of Defense shall be used to—(1) retire, divest, realign, or transfer aircraft of the Air Force; (2) disestablish or convert any unit associated with aircraft described in paragraph (1) or any unit of the Air National Guard or Air Force Reserve; or (3) retire C-23 Sherpa aircraft." (§ 121)
2014	113-39	September 30, 2013	None	The sole purpose of this CR, the Pay Our Military Act, CR was to provide funds to pay DoD military, civilians, and contractors.
2014	113-44	October 10, 2013	December 15, 2013	The sole purpose of this CR, the Department of Defense Survivor Benefits Continuing Appropriations Resolution, 2014, was to provide funds to pay DoD military, civilians, and contractors.
2014	113-46	October 17, 2013	January 15, 2013	None
2014	113-73	January 15, 2013	January 18, 2013	None
2015	113-164	September 19, 2014	December 11, 2014	"Notwithstanding any other provision of law, except sections 106 and 107 of this joint resolution, for 'Department of Defense—Overseas Contingency Operations—Operation and Maintenance—Operation and Maintenance, Army', up to $50,000,000, to be derived by reducing the amount otherwise made available by section 101 for such account, may be used to conduct surface and subsurface clearance of unexploded ordnance at closed training ranges used by the Armed Forces of the United States in Afghanistan . . . ." (§ 119) "The Department of Defense may submit a reprogramming or transfer request to the congressional defense committees for funds made available by section 101(a)(3) of this joint resolution and designated in section 114 of this joint resolution to carry out activities authorized under this section notwithstanding sections 102 and 104 of this joint resolution." (§ 149[f]) "The Secretary of Defense is authorized, in coordination with the Secretary of State, to provide assistance, including training, equipment, supplies, and sustainment, to appropriately vetted elements of the Syrian opposition and other appropriately vetted Syrian groups and individuals . . . ." (§ 149[a])
2015	113-202	December 12, 2014	December 13, 2014	None

Table 1.3—Continued

FY	Public Law	Enactment Date	Expiration Date	DoD Anomaly
2015	113-203	December 13, 2014	December 17, 2014	None
2016	114-53	September 30, 2015	December 11, 2015	None
2016	114-96	December 11, 2015	December 16, 2015	None
2016	114-100	December 16, 2015	December 22, 2015	None
2017	114-223	September 29, 2016	December 9, 2016	"[T]he Secretary of Defense may develop, replace, and sustain Federal Government security and suitability background investigation information technology system requirements of the Office of Personnel Management at a rate for operations of $95,000,000." (§ 121)
2017	114-254	December 12, 2016	April 28, 2017	"Notwithstanding sections 101, 102, and 104 of this Act, from within amounts provided for 'Department of Defense—Procurement—Shipbuilding and Conversion, Navy', funds are provided for 'Ohio Replacement Submarine (AP)' at a rate for operations of $773,138,000." (§ 101) "Notwithstanding sections 102 and 104 of this Act, amounts made available pursuant to section 101 may be used for multiyear procurement contracts, including advance procurement, for the AH-64E Attack Helicopter and the UH-60M Black Hawk Helicopter." (§ 101) "Notwithstanding section 102, funds made available pursuant to section 101 for 'Department of Defense—Procurement—Aircraft Procurement, Air Force' are provided for the KC-46A Tanker up to the rate for operations necessary to support the production rate specified in the President's fiscal year 2017 budget request." (§ 101) And • funding for ERI across military personnel and O&M accounts • funding for counterterrorism operations across military personnel, O&M, procurement, and RDT&E accounts • funding for the Iraq Train and Equip Fund • funding for the Joint Improvised-Threat Defeat Organization to support counterterrorism operations
2017	115-30	April 28, 2017	May 5, 2017	None

SOURCE: U.S. Congress.

NOTE: ERI = European Reassurance Initiative. O&M = operation and maintenance. RDT&E = research, development, test, and evaluation.

operational.[15] The 178-day CR opening FY 2013 came with several anomalies, reflective of the challenge of operating under a long-duration CR. The CRs passed at the beginning of FY 2014 (when the government was in a two-week-long shutdown) provided anomalies to support sensitive priorities of military pay and payments for survivor benefits. One anomaly provided for in the three-month-long CR opening FY 2015 supported operations in Syria. The policy change ("new start") regarding the nature of U.S. engagement in Syria apparently could not have been resourced under a CR in the absence of this anomaly.

In contrast to those for previous FYs, many requests for FY 2017 made it into the legislation as anomalies. Anomalies in this FY addressed resources for the Office of Personnel Management hack; provided resources for the Ohio Replacement Submarine and the KC-46A, as well as MYP authority for the AH-64E Attack Helicopter and the UH-60M Black Hawk Helicopter; and supported operations in Europe, Syria, and Iraq. With respect to the anomaly request for ERI, the Army Secretary reported in September 2016, ERI "funds increase pretty substantially from this year to last," so being denied an anomaly would hamper the military's ability to ramp up deterrence activities.[16]

Information on anomaly requests included in the list in Table 1.3 of CR legislation provides insight into a tool that DoD uses to attempt to maintain a degree of flexibility under a CR. Although, to avoid substantial disruption of its planned programs, DoD frequently asserts the need for exceptions to CR-imposed constraints, relatively few provisions eventually appear in legislation. The greater number of anomalies provided for in the FY 2017 CR reportedly reflects the extent to which many observers anticipated that it could be a long time before a defense appropriation would be passed.[17]

Recent Practice

As noted earlier and displayed in Table 1.1, federal operations under a CR have become the norm in recent years. More specifically, in six of the years since FY 2009, none of the 12 appropriation bills was enacted by the start of the FY.[18] As can be seen in Table 1.1, this has meant that, in the absence of an appropriation, Congress has been required to pass several short-term CRs to sustain government operations. The CRs listed in Table 1.1 are only those affecting DoD and are a subset of CRs passed in most

---

[15] Brennan et al., 2013, p. 117.

[16] Jen Judson, "Fanning: Continuing Resolution Could Snarl European Reassurance Initiative Efforts," *Defense-News*, October 2, 2016.

[17] A defense appropriation for FY 2017 was passed in May 2017.

[18] Saturno and Tollestrup, 2016, Table 1.

years.[19] Also, amid the FY 2014 government shutdown halting many federal operations, Congress did pass two CRs affecting DoD in a limited manner: one providing pay and allowance for military personnel and civilians and contractors supporting those personnel and a second providing for the payment of death gratuities and other funeral expenses for military personnel.[20]

It is important to note that, in recent years, DoD has been less affected by CRs than other federal departments and agencies have, although it has still gone without regular appropriations for many weeks or several months. This has meant that DoD operates under CRs for shorter periods of time before receiving a regular appropriation than other elements of the federal government do.[21] However, Congressional Research Service (CRS) authors also note that the amount of time DoD has operated under CRs has increased in recent years.[22]

Consequences of Operating Under a Continuing Resolution

Although a CR is intended to preserve congressional prerogatives to make final decisions on full-year funding levels and to prevent a funding gap and government shutdown, operating under a CR is distinct in important ways from operating under a regular appropriation and is widely thought to have negative consequences for resource management.

---

[19] Others affect other agencies. For example, in Public Law 113-6, Consolidated and Further Continuing Appropriations Act, 2013, March 26, 2013, Division F provided continuing appropriations for FY 2013 for the Energy and Water Development and Related Agencies Appropriations Act, 2012 (Pub. L. 112-74, Consolidated Appropriations Act, 2012, December 23, 2011, Division B); the Financial Services and General Government Appropriations Act, 2012 (Pub. L. 112-74, 2011, Division C); the Department of the Interior, Environment, and Related Agencies Appropriations Act, 2012, (Pub. L. 112-74, 2011, Division E); the Departments of Labor, Health and Human Services, and Education, and Related Agencies Appropriations Act, 2012 (Pub. L. 112-74, 2011, Division F); the Legislative Branch Appropriations Act, 2012 (Pub. L. 112-74, 2011, Division G); the Department of State, Foreign Operations, and Related Programs Appropriations Act, 2012 (Pub. L. 112-74, 2011, Division I); and the Transportation, Housing and Urban Development, and Related Agencies Appropriations Act, 2012 (Pub. L. 112-55, Division C). In Public Law 113-6, Divisions A through E contain the texts of the Agriculture, Rural Development, Food and Drug Administration, and Related Agencies Appropriations Act; Commerce, Justice, Science, and Related Agencies Appropriations Act; the Department of Defense Appropriations Act; Department of Homeland Security Appropriations Act; and Military Construction and Veterans Affairs and Related Agencies Appropriations Act. We do not list Public Law 113-6 as a CR affecting DoD because, although it specified continuing appropriations for several agencies, it also contains the FY 2013 Defense Appropriations Act. See Saturno and Tollestrup, 2016, Table 3, pp. 17–18.

[20] Amy Belasco and Pat Towell, *Government Shutdown: Operations of the Department of Defense During a Lapse in Appropriations*, Washington, D.C.: Congressional Research Service, R41745, October 15, 2013.

[21] Belasco and Towell, 2013.

[22] Williams and Wees, 2016, summary.

In this section, we review the literature discussing consequences of operating under a CR, as well as the cases made by government officials regarding those consequences. It is important to note that we have not independently assessed the validity of the basis for the consequences attached to these cases and do not endorse those consequences; rather, we simply report them. Notably, research for the literature described draws primarily on insights from interviews with personnel responsible for managing organizations operating under CRs or personnel with expertise in budget dynamics.[23] In these accounts, the scale and nature of potential deleterious effects anticipated from operating under a CR are self-reported and anecdotal, which makes drawing broad conclusions challenging (if not impossible) because data enabling definitive conclusions that the effects actually occurred are lacking. Nonetheless, a review of the literature is useful for establishing the breadth of effects observers often assert are associated with CRs.

Inefficiencies

Perhaps the most widely presumed cost associated with operating under a CR is inefficiency, or missed opportunities for cost avoidance, which suggests that a CR can make both processes for personnel management and procurement less efficient.

With respect to personnel management, several researchers and observers have identified dynamics of operating under a CR with potential for both near-term and long-term deleterious effects. In the near term, this work suggests, a CR could result in a mismatch among staffing levels, planned or mandated activities, and available funding. For example, a given agency might be forced to impose a hiring freeze or a delay in hiring, or managers might opt to leave positions unfilled to hedge against future funding shortfalls or in anticipation of pay increases.[24] In 2017 congressional testimony, Chief of Staff of the Army Mark A. Milley indicated that such dynamics "significantly risks a return to a hollow Army," in that "[m]andated end strength without commensurate funding will mean [that] only a select few units will be ready for combat."[25]

---

[23] Philip G. Joyce, *The Costs of Budget Uncertainty: Analyzing the Impact of Late Appropriations*, Washington, D.C.: IBM Center for the Business of Government, Improving Performance series, 2012; GAO, *Continuing Resolutions: Uncertainty Limited Management Options and Increased Workload in Selected Agencies—Report to the Ranking Member, Subcommittee on Oversight of Government Management, the Federal Workforce, and the District of Columbia, Committee on Homeland Security and Governmental Affairs, U.S. Senate*, Washington, D.C., GAO-09-879, September 2009; Michelle Sager, director, strategic issues, U.S. Government Accountability Office, *Budget Issues: Effects of Budget Uncertainty from Continuing Resolutions on Agency Operations—Statement for the Record by Michelle Sager, Director, Strategic Issues*, Washington, D.C.: U.S. Government Accountability Office, GAO-13-464T, March 13, 2013.

[24] GAO, 2009; Clinton T. Brass, *Interim Continuing Resolutions (CRs): Potential Impacts on Agency Operations*, Washington, D.C.: Congressional Research Service, RL34700, March 16, 2010; Samuelsohn, 2015; Joyce, 2012; Sager, 2013.

[25] GEN Mark A. Milley, Chief of Staff, U.S. Army, "Record Version: Statement by General Mark A. Milley, Chief of Staff, United States Army, Before the Committee on Armed Services, U.S. House of Representatives,

For the staff members at a given agency, a CR might also affect professional development and morale. For example, a CR can curtail activities, such as travel or training, and halt compensation, such as overtime pay.[26] The result might be negative effects on morale and, ultimately, on recruitment and retention.[27] A factor exacerbating these dynamics that some researchers have suggested is the fact that operating under a CR can carry additional workload for staff. Additional work, such as planning, issuing guidance, monitoring resources, and reporting to Congress and OMB, might be associated with managing the organization's response to the CR.[28] For example, for an O&M contract, the contract value for a specific piece of work might have to be negotiated more than once, under each CR increment. Also, additional workload might be associated with the CR because of repetitive tasks that could have been avoided under the provisions of a regular appropriation. For example, an interim CR can necessitate short-term contracts that must be reissued once additional funding is provided, additional paperwork, and additional overhead in contracting actions.[29] Such repetitive work and increased transaction costs might also increase the risk of error, audit, or waste.[30] These researchers suggest that the risk of understaffing, coupled with additional workload and policies that erode employee morale, could lead to reduced productivity and less work completed.

Other literature indicates that a CR might negatively affect the efficiency with which the government procures goods and services. This literature suggests that a CR increases programmatic risk, especially with respect to cost and schedule, through several mechanisms.[31] First, a CR's prohibition on funding for activities that did not receive an appropriation the previous year might prevent a program manager from letting contracts on the timeline planned.[32] For example, Northrop Grumman's chief executive officer gave voice in 2017 to a widely held belief that a CR "certainly does impact the timing of the customer's ability to get out there and make source selection decisions."[33] In 2017 congressional testimony, Chief of Staff of the Air Force David L. Goldfein indicated that a CR "[r]estricts our ability to award the Long Range Stand-

First Session, 115th Congress, on Consequences to the Military of a Continuing Resolution," Washington, D.C.: U.S. House of Representatives, April 5, 2017.

[26] Joyce, 2012; Brass, 2010.

[27] Joyce, 2012.

[28] Brass, 2010; Sager, 2013; GAO, 2009.

[29] Williams and Wees, 2016, summary; Samuelsohn, 2015.

[30] GAO, 2009; Joyce, 2012.

[31] Joyce, 2012.

[32] Brass, 2010; Samuelsohn, 2015.

[33] Marjorie Censer, "Contractors Remain Upbeat on Defense Spending—but Warn of Uncertainty," *Inside the Navy*, Vol. 30, No. 17, May 1, 2017.

off Weapon and Ground Based Strategic Deterrent technology-maturation and risk-reduction contracts on time."[34]

Second, observers also note that the constraints of operating under a CR can hinder the efficient evolution of an acquisition program. As CRS researchers observed,

> if a CR provides funds at the rate of the prior year's appropriation, an agency may be provided additional (even unneeded) funds in one account, such as research and development, while leaving another account, such as procurement, underfunded or lacking. This is sometimes referred to as a problem with the color of money.[35]

An inability to move a program forward also could incur costs associated with sustaining legacy programs beyond their service lives, some have argued. For example, Air Force senior leaders have maintained that such additional costs resulted in inefficiencies in the next-generation bomber, intercontinental ballistic missile, and EC-130H Compass Call programs.[36] The evolution of an acquisition program might also be hampered by an inability under a CR to increase production rates. For example, the Commandant of the Marine Corps observed, "FY17 CR based prohibitions on program new starts and quantity increases, as well as limitations on investment funding at line item levels, will prevent planned funding and production rate increases for multiple ground and aviation programs."[37]

Finally, the literature also suggests that a CR might prevent program managers from taking advantage of more cost-effective acquisition strategies. CRS analysts explained that, under a CR, "[i]t is also likely that DoD would be limited in its ability to enter into planned long-term contracts, thus losing the program stability and efficiencies that can be gained by such contracts."[38] One example of a more efficient acquisition approach that might be hampered under a CR is MYP.[39] Congress must

---

[34] Gen. David L. Goldfein, Chief of Staff, U.S. Air Force, "Department of the Air Force Presentation to the Committee on Armed Services, United States Congress; Subject: Impacts of a Year-Long Continuing Resolution; Statement of: General David L. Goldfein, Chief of Staff of the Air Force," Washington, D.C.: U.S. House of Representatives, April 5, 2017.

[35] Williams and Wees, 2016, summary; see also Dana A. Shea and Daniel Morgan, *Federal Research and Development Funding: Possible Impacts of Operating Under a Continuing Resolution*, Washington, D.C.: Congressional Research Service, RS22978, October 27, 2008.

[36] Rachel S. Karas, "Long-Term CR Could Delay Compass Call Fielding by at Least a Year," *Inside the Air Force*, Vol. 28, No. 17, April 28, 2017; Goldfein, 2017.

[37] Gen Robert B. Neller, Commandant, U.S. Marine Corps, "Statement of General Robert B. Neller, Commandant of the Marine Corps, Before the House Armed Services Committee on Continuing Resolution Consequences," Washington, D.C.: U.S. House of Representatives, April 5, 2017.

[38] Williams and Wees, 2016, p. 6.

[39] For additional discussion, see Ronald O'Rourke and Moshe Schwartz, *Multiyear Procurement (MYP) and Block Buy Contracting in Defense Acquisition: Background and Issues for Congress*, Washington, D.C.: Congres-

authorize the use of a multiyear contract based on several criteria and requirements.[40] Annual appropriation processes allow Congress flexibility in providing oversight and for adjusting funding in accordance with changes to the program or strategic environment. If granted MYP authority, DoD can use a single contract to cover procurement costs over two to five years, rather than year to year. This approach has been found under some circumstances to reduce procurement costs due to efficiencies from stability in contractor workforce and plans, production and supply efficiencies, and reduced administrative costs, but it might be unavailable to a manager operating under a CR.[41]

Constraint on Exercise of Managerial Discretion

The literature also reflects a concern that operating under a CR limits senior leaders' ability to plan or set priorities. This means that a CR could hamper senior leaders' ability to respond to changes and adjust to dynamic situations. One journalistic account relying on interviews with senior budget officials noted that "[o]perating during a CR, or under the perpetual threat of one, leads to hesitant, backward-looking thinking that would never be duplicated in the business world."[42] Then–Defense Secretary Ashton B. Carter echoed this sentiment when he asserted that CRs put commanders in a "straight-jacket" that hinders the Pentagon's ability to adapt to meet dynamic national security challenges.[43] General Milley's 2017 testimony similarly emphasized that, in the face of the dynamic change that defines the future battlefield, a CR was an inadequate tool for resource allocation.[44] He asserted that a CR "will force the Army to defer and cancel modernization efforts across both our air and ground fleets that address immediate capability gaps and build our future Army."[45] In addition to the extent to which a CR

---

sional Research Service, R41909, March 23, 2016. The appendix to that report lists programs approved for MYP from FY 1990 to FY 2016.

40 For a summary of MYP requirements and criteria, see Obaid Younossi, Mark V. Arena, Kevin Brancato, John Graser, Benjamin W. Goldsmith, Mark A. Lorell, Fred Timson, and Jerry Sollinger, *F-22A Multiyear Procurement Program: An Assessment of Cost Savings*, Santa Monica, Calif.: RAND Corporation, MG-664-OSD, 2007, pp. 9–18, and GAO, *Defense Acquisitions: DoD's Practices and Processes for Multiyear Procurement Should Be Improved*, Washington, D.C., GAO-08-298, February 7, 2008.

41 CRS researchers reported that estimated development cost savings ranged from 5 to 10 percent (O'Rourke and Schwartz, 2016). The Office of the Secretary of Defense's office of Cost Assessment and Program Evaluation analysis of four aircraft procurement programs (F/A-18E/F strike fighters, H-60 helicopters, V-22 tilt-rotor aircraft, and CH-47F) estimated cost savings to range from 2 to 8 percent (O'Rourke and Schwartz, 2016, pp. 3–4). RAND analysis of MYP of the F-22 estimated a cost savings of $411 million, or about a savings of 4.5 percent, lower than historical range estimates (Younossi et al., 2007). However, GAO questioned the strength of the empirical evidence on cost savings supporting MYP decisions (GAO, 2008).

42 Samuelsohn, 2015.

43 Williams and Wees, 2016, p. 5.

44 Milley, 2017.

45 Milley, 2017.

has been perceived as hampering managers' ability to plan, observers have argued, a CR could affect DoD's ability to prioritize. For example, CRS authors indicated that, although the FY 2017 President's budget called for prioritizing readiness (an $8.4 billion increase for O&M relative to its FY 2016 appropriation), the CR constrained senior leaders' ability to reflect this priority in increased resource levels.[46]

An Approach for Identifying One Type of Costs Associated with Operations Under a Continuing Resolution

Because of a lack of quantitative data, many of the consequences discussed in the previous section would be very difficult to estimate quantitatively or to conclusively demonstrate. All of the research that we reviewed on the consequences of operating under a CR employed qualitative approaches that focused on case studies, assertions, and anecdotal information. For example, the most substantial work in this area was done by GAO for studies in 2009 and 2013 that considered the sources of additional workload associated with operating under a CR.[47] GAO's case-study approach provided insight by means of self-reported estimates on a diverse range of issues. By the authors' own admission, this approach had limitations, however. Although the research team took steps to support standardization and comparability across its case studies, it was impossible to remove several sources of variation. For example, the authors noted,

> One of the limitations of our case study analysis is that we had to rely to a large degree on testimonial evidence because case study agencies could not provide documentation showing the foregone [sic] opportunities resulting from a CR. In general, agencies do not produce planning documents—such as spending plans or monthly hiring targets—until they have received their regular appropriations. Aside from [the U.S. Department of Veterans Affairs], all case study agencies have operated under a CR for each of the past 11 years and therefore could only speculate on how they would have operated differently or more efficiently, except anecdotally. In general, there were too many variables for agencies to isolate the effects of CRs from other factors.[48]

As a result of such limitations, GAO acknowledged that, although "selected case studies cannot be generalized," the research did suggest "broad-based commonalities

---

46 Williams and Wees, 2016, summary.

47 Case studies included were U.S. Department of Health and Human Services components the Administration for Children and Families and the U.S. Food and Drug Administration, U.S. Department of Veterans Affairs components the Veterans Health Administration and Veterans Benefits Administration, and U.S. Department of Justice components the Federal Bureau of Prisons and Federal Bureau of Investigation (GAO, 2009; Sager, 2013).

48 GAO, 2009, p. 32.

in the experiences of federal agencies."[49] However, GAO was unable to estimate the costs beyond what any agency's operations would have been in the absence of a CR. In short, a CR's effects on a given agency are a function of many variables, including the nature of funding, nature of service provided, management decisions, congressional action, prior-year funding, and year-to-year dynamism. In this context, a qualitative case-study approach provides some insight into a policy question about which it is difficult to perform quantitative analysis yielding defensible, generalizable results.

Although we still confront the problem that many variables can mask CRs' effects, we adopted a different approach for this study. We focused our analysis on one widely cited effect of a CR—a constraint on the ability to make procurement awards—and we limited the scope of our analysis to DoD procurement programs. We considered contract actions for relatively high-profile programs, and we looked for indication that a CR could be associated with contract actions that were delayed or unit costs that increased relative to plans. In particular, for these contract actions, we compared originally planned and actually achieved award dates, as well as originally estimated and actually achieved unit costs and procurement quantities. We performed statistical analysis to determine whether there were significant differences in these metrics between awards originally planned during periods when CRs subsequently occurred, versus awards planned during periods that did not experience a CR.[50] We also made a limited attempt to determine whether CRs' systematic effects on contract awards might be occurring by comparing outcomes for awards planned for FY 1999 (three weeks of operations under CRs) with outcomes for awards planned for FY 2013 through FY 2015 (operations under CRs lasting multiple months).[51]

---

[49] GAO, 2009, p. 32.

[50] Note that our statistical analysis cannot definitively link cause and effect. In our analysis, awards are planned to occur either during a CR or not and are assigned to one or the other of these binary categories based on the data available in DoD's budget materials, which most often specify only the month an award is planned, not the precise day. The last FY 2013 CR extended through March 27, 2013, and covered 19 of 21 business days in March (or 90 percent). The last FY 2014 CR extended through January 18, 2014, and covered 12 of 21 business days in January (or 57 percent). The last FY 2015 CR extended through December 17, 2014, and covered 13 of 22 business days in December (59 percent). Therefore, given that each of the CRs covered the majority of the business days during the last month it was in effect, and in the absence of readily available precise award dates, we consider it reasonable to assign to the during-a-CR category all awards planned to occur during the final month of the CRs we considered. This assumption applies only to our analysis of outcomes associated with awards during and not during CRs planned during FYs 2013 through 2015.

[51] This comparison has limitations. For example, comparing FY 1999 awards with FY 2013 and FY 2014 awards still leaves open the question of what effects legislative changes, such as the Weapon Systems Acquisition Reform Act of 2009 (Pub. L. 111-23, 2009) (WSARA), and the many changes that have occurred since 1999 (and continue to occur) in the Federal Acquisition Regulation, might be having.

We employed four complementary sources of data for our analysis of contract actions:

- First, we reviewed DoD-submitted requests for anomalies for procurement programs. These requests provided insight into specific programs that the Pentagon identified as at risk of suffering particularly deleterious consequences under a CR.
- Second, we used the budget justification materials—in particular, the procurement exhibits, including the P-5a forms providing information on contract award dates and unit costs, across the FYs (2013 through 2015 and 1999 through 2000) that were the focus of our analysis. We assessed 199 programs, the majority of which are major acquisition programs, including those cited in DoD's anomaly requests.[52]
- Third, when the DoD budget justification materials appeared inconsistent or unclear, we looked to an alternative source of information on contract actions: USAspending.gov.[53]
- Fourth, we considered the Selected Acquisition Reports (SARs) for narrative descriptions of the sources of programmatic issues and changes.[54]

We compared projected versus realized plans for 151 procurement awards (consisting of a mix of awards under new and existing contracts) for Army, Navy, Marine Corps, and Air Force programs as documented in the services' President's budget submissions for FYs 2013 through 2015 (see the appendix). We focused our assessment on these years, particularly the first two, because, until this past year, two of the longer periods during which the government operated under CRs were at the beginnings of FYs 2013 and 2014 (see Table 1.1).[55] However, to gain insight into whether CRs might be having systematic effects on all planned awards, we also compared the outcomes of

---

[52] Office of the Under Secretary of Defense (Comptroller), "DoD Budget Request," undated. We assessed 199 procurement contract awards, the majority of which were associated with major acquisition programs. Of those, 151 were planned or made between FY 2013 and FY 2015, and 48 were planned initially in FY 1999. The appendix lists the programs associated with those awards. The number of awards is greater than the number of programs because we consider multiple awards proposed or made in each program across the years.

[53] USAspending.gov, home page, undated (a).

[54] A SAR provides "a comprehensive summary of a Major Defense Acquisition Program (MDAP) Acquisition Category (ACAT) I program that is required for periodic submission to Congress by the Secretary of Defense. It's mandated by Title 10 USC § 2432 'Selected Acquisition Reports'" ("Acquisition Process: Selected Acquisition Report (SAR)," *AcqNotes*, updated July 11, 2017). Some SARs are publicly available (Office of the Secretary of Defense and Joint Staff, "Office of the Secretary of Defense and Joint Staff, Freedom of Information Act Requester Service Center," undated). We used the unclassified SARs for FYs 2012 through 2016 to obtain admittedly limited information regarding selected aspects of the execution of the programs considered in our analysis. We list those programs in the appendix.

[55] We could not perform a definitive analysis of awards for FY 2016 or FY 2017 procurements because insufficient time has passed for the limited available documentation to provide definitive information on projected versus actual awards for FY 2016 or FY 2017.

the FY 2013 through FY 2015 awards with 48 awards planned initially to occur during FY 1999, which was affected by CRs for only three weeks. In the "Results" section, we discuss our observations based on our analysis of these data.

Limitations of Our Approach

Our approach gives us insight into only a limited portion of the larger story about how CRs might affect DoD's decisionmaking and operations. As we stated at the outset, this focused, preliminary effort is not a comprehensive analysis of the many issues of potential interest associated with DoD operations under a CR, and even this focused effort must contend with limitations in the available data. For example, we were unable, because of a lack of readily available data in the budget materials and SARs, to determine, for each case in which an award is made, the FY in which the funds used to make the award were appropriated. (Prior-year funds can be used to make awards in a new FY during a CR.)

It was also beyond the scope of our focused effort to comprehensively address or attempt to systematically control for congressional marks modifying the President's budget request other than to observe, as we have, the case in which we found information that cites these marks as having had deleterious effects.

Given the limited duration of the time periods we analyzed, our analysis also did not address the potential effects of sharp, sustained increases and decreases to total DoD funding (the DoD top line), which would require comparing outcomes during periods of sharp increases and decreases with periods in which top-line budgets were not changing rapidly (the latter being the case for our analysis, notwithstanding sequestration). Similarly, we did not seek to understand what effects, if any, changes in the National Military Strategy might have on outcomes, which would require comparing outcomes among multiple periods that adopted markedly different strategies. (We note however, that the National Military Strategy did not change significantly between FY 2013 and FY 2015 or between FY 1999 and FY 2000, although it did change, arguably significantly, between the two periods.)

With few exceptions, we focused on Acquisition Category I and II programs, not seeking to assess whether there are differences in outcomes between these two program categories or between major and nonmajor programs. We also could not identify other potential effects, such as changes in program management or in acquisition laws and regulations. Data permitting the former to be assessed comprehensively are not readily available, if available at all. (However, we expect that, by considering a relatively large number of procurement awards across the services, as we have done for the years on which we focused, the fraction of awards made in programs experiencing management changes will be small.) Regarding the latter, there have been steady changes in the acquisition laws and regulations, with occasional, more-substantial changes, such as those associated with WSARA. Developing and applying methods that might be used to control for those steady and abrupt changes, which would, in particular, entail developing some means of evaluating the significance of the steady changes and when that significance has exceeded some threshold, is beyond the scope of this limited effort.

Our data set also did not include a substantial number of procurement awards planned under new contracts and, in particular, because it focused on procurement contracting, excluded awards for RDT&E efforts, including new-start RDT&E contracts.

Finally, in some cases, we had sufficient data points to ascertain statistical significance only for relatively large effects (i.e., only for relatively large differences in the quantitative metrics that we considered, such as delays in procurement award dates). Thus, there could be smaller significant differences in these metrics that could be discovered only by analysis of a larger set of data than used in our limited analysis.

Results

Our analysis of DoD's projected versus realized procurement awards does not provide strong evidence that CRs are associated with the following problems that DoD cites as possibilities in the anomalies it proposes:[56]

- delays in procurement that, in turn, delay fielding needed capabilities
- delays in contract awards causing production breaks, production-line requalification, or cost increases
- broken MYPs and cost increases.

Nonetheless, because of the limitations discussed, our analysis also did not definitively demonstrate the absence of these effects.

We first discuss the results of our analysis of only those awards planned originally to occur during FY 2013 through FY 2015. Because CRs of multiweek, if not multimonth, duration occurred during each of these FYs, the results of this analysis could include systematic effects associated with CRs, such as program managers anticipating the occurrence of a CR each year and adjusting their planning accordingly.

In an attempt to control for this kind of systematic effect, we have analyzed the differences between outcomes for awards planned for FYs 2013 through 2015 and awards planned for FY 1999, when CRs spanned only three weeks. Although CRs occurred regularly during the ten-year periods preceding both FY 1999 and FY 2015,

[56] As discussed already, Congress has ultimately included in CRs few, if any, of the draft anomalies DoD has prepared. We obtained from the staff of the Office of the Under Secretary of Defense for Acquisition, Technology and Logistics draft anomalies prepared within DoD for FYs 2013 through 2017 likely constituting a subset of all the anomalies prepared and considered in DoD for those years. Those anomalies address potential CR-related problems in all of DoD's activities. Our analysis addressed only CRs' potential effects on procurement awards made using new and existing contracts for weapon systems.

their average durations were significantly different: 28 days between FY 1990 and FY 1999 and 86 days between FY 2006 and FY 2015 ($p = 0.0169$). This difference suggests that, if program managers were anticipating CRs throughout the period from FY 1999 to FY 2015, they should have been anticipating much shorter CRs and therefore been making less significant adjustments to their plans in FY 1999 than between FY 2013 and FY 2015.

Finally, we consider the specifics of several awards that DoD-prepared draft anomalies indicated would be delayed with concomitant negative effects as a result of CRs.

Analysis of Awards for Fiscal Years 2013 Through 2015

For the set of 151 procurement awards planned for FY 2013 through FY 2015 that we considered, our analysis indicates the following (see Tables 1.4 through 1.8):

- Irrespective of when awards were originally planned, delays occurred about as often as no delays or accelerations.[57]
- In the proportion of awards that experienced a delay or in the lengths of the delays experienced, we found no statistically significant difference overall between awards planned originally to occur during a period when a CR subsequently occurred and awards planned later in the year after an appropriation bill was enacted.[58]
- In some cases, awards originally projected to occur outside a CR period were subsequently made during a CR, and, overall, a substantial number of awards, including for MYPs and procurements using new contracts, were made during periods when CRs were in effect.
- Whenever awards were made, decreases or no change in unit costs relative to original projections outnumbered increases.
- Projected versus realized unit costs declined slightly on average for all FY 2013–2015 awards, and there was not a statistically significant difference in unit costs between awards planned and not planned for periods subsequently affected by a CR.
- Whenever awards were made, increases or no change in numbers of items procured relative to original projections outnumbered decreases.

The appendix lists the awards and associated programs we considered. We selected these awards with no prior knowledge of the details of their contractual histories. Although resource limitations precluded consideration of all procurement awards

[57] For the originally projected date, we used the date for contract award appearing for the first time in the President's budget submitted for the FY in which the award was to occur; we treated contract unit cost similarly.

[58] We calculated net changes in award dates for awards made later than original projections (delays) and for awards made earlier than original projections (accelerations); we calculated delays using only the former.

projected for the three-year period, our assessment generally included more than one-half of the procurement awards that each of the services planned for major defense programs, as well as for several other, lesser programs, for FYs 2013 and 2014.[59]

Table 1.4 displays selected descriptive metrics for the 151 procurement awards planned for FY 2013 through FY 2015 that we considered in this study. Although not supporting definitive conclusions because they consider only awards planned for FY 2013 through FY 2015, which might be suffering from systematic effects of CRs, the results displayed in Table 1.4 suggest, in some cases, that program managers might not all now be developing their plans assuming that CRs will occur: Sixty-two percent of award dates initially planned would have been made during the CRs that subsequently occurred, notwithstanding the knowledge that, for many years, CRs have regularly spanned multiple months at the beginning of each new FY. The results displayed in Table 1.4 also suggest that CRs might not systematically be causing all awards to be delayed: Odds are almost even that an award will proceed as planned or be accelerated rather than be delayed. Moreover, the results suggest that CRs might not be causing all costs to increase: Reductions in originally projected unit costs occurred more than twice as often as cost increases did. On the other hand, this could also indicate that program managers are being conservative in making their original cost estimates, which could be due to their anticipation of CRs but could also be due to the changes in acquisition laws and regulations, such as the enactment in FY 2009 of WSARA, that, among other provisions, put greater emphasis on better cost-estimating.

Table 1.5 considers the average change between original projected award date and the actual award date for awards originally projected to occur during and not during a subsequent CR. We found the difference between the average net change in awards (in days, or award-days) planned to occur during CRs to be significantly longer than the change in days for awards not planned to occur during CR periods: 120 days versus 54 days. Nonetheless, the proportion of awards experiencing a delay (of any length) in each category (planned during a CR versus not planned during a CR) is

59 For example, for FY 2014, our assessment included 13 of 15 Air Force major programs with awards planned, 23 of 40 major Navy programs with awards planned, and 17 of 24 major Army programs with awards planned. In this analysis, *major program* refers to a defense acquisition program that meets the statutory definition (see U.S. Code, Title 10, Armed Forces, Subtitle A, General Military Law, Part IV, Service, Supply, and Procurement, Chapter 144, Major Defense Acquisition Programs, Section 2430, Major Defense Acquisition Program Defined), but it also includes selected other programs (e.g., awards for programs that do not meet the statutory definition but are mentioned in DoD proposals for anomalies). In some instances, multiple awards were planned per program (e.g., for aircraft programs, separate awards can be planned for airframes and engines). In such cases, our assessment usually included one planned award (e.g., for the airframe) per program. Because contract awards for all three services' annual procurements are made simultaneously, we included Joint Strike Fighter (JSF) procurement awards planned by the Air Force but not those planned by the Navy (including the Marine Corps). We also generally excluded procurements associated with programs that are known to have undergone substantial revisions to original plans to incorporate changes in procurement plans clearly not related to CRs, such as the Army's planned purchases of radios developed under the Joint Tactical Radio System.

Table 1.4
Metrics for Selected Defense Procurement Awards Projected to Occur Between Fiscal Year 2013 and Fiscal Year 2015

Metric	Value	Percentage of Total
Procurement awards considered	151	
Awards delayed relative to original projections	78	52
Awards accelerated relative to original projections	23	15
Awards with no change in schedule relative to original projections	50	34
Awards with unit-cost increases relative to original projections	38	24
Awards with unit-cost decreases relative to original projections	86	58
Awards with no change in unit cost relative to original projections	27	18
Awards with quantity decreases relative to original projections	32	21
Awards with quantity increases relative to original projections	43	28
Awards with no change in quantity relative to original projections	76	50
Awards originally projected to occur during CRs	93	62
Awards that actually occurred during CRs	71	47
Likelihood of delay versus no change or acceleration[a]	1.1	
Likelihood of unit-cost increases versus decreases or no change[a]	0.3	
Likelihood of quantity decreases versus increases or no change[a]	0.3	

SOURCE: DoD data, including budget exhibits and SARs for selected programs listed in the appendix.

NOTE: Because of rounding, percentages in each section do not sum to 100.

[a] We calculated likelihood of delay versus no change or acceleration as the number of programs examined that experienced a delay in the originally projected contract award divided by the number of programs that experienced no delay or an award made earlier than the originally projected date. The other likelihoods displayed are calculated analogously.

not significantly different.[60] (And, given their contents, there is no apparent reason that programs experiencing delays would be considered lower priority than those not experiencing delays.) This apparent inconsistency can be explained by the fact that the average change in award-days explicitly includes negative values for those cases in which awards were made sooner than originally planned (i.e., accelerated). However, awards planned later in an FY can be accelerated by a greater number of days than

[60] Of the 151 awards planned for FYs 2013 through 2015, 93 were planned to occur during subsequent CRs, and 53 of those awards experienced a delay. Of the 58 awards not planned to occur during subsequent CRs, 25 experienced a delay. A Fisher two-tailed test on these data yields $p = 0.1316$, indicating no statistically significant difference between the two cases in the proportion of awards experiencing a delay.

Table 1.5
Change in Award Dates for Procurement Awards Projected Originally to Occur During and Not During a Continuing Resolution for Fiscal Years 2013 Through 2015

Metric	Value, in Days
Net change overall	
All services, programs, during	120
All services, programs, not during[a]	54
All services, greatest delay, during	639
All services, greatest acceleration, during	–90
All services, greatest delay, not during	579
All services, greatest acceleration, not during	–274
Delays only	
All services, during	219
All services, not during[b]	189
Average durations of FY 2013–2015 CRs	
Duration of FY 2013 CRs	177
Duration of FY 2014 CRs	109
Duration of FY 2015 CRs	67

SOURCE: DoD data, including budget exhibits and SARs for the programs listed in the appendix.

[a] Difference between during and not during is significant ($p = 0.0155$); effect size = 0.4096; 95 percent confidence interval for effect size = [0.0784, 0.7408].

[b] Difference between during and not during is not significant ($p = 0.4108$); effect size = 0.4096; 95-percent confidence interval for effect size = [–0.2759, 0.6773].

awards planned earlier in a FY, which is when CRs occur. This is reflected in the 274-day greatest acceleration displayed in Table 1.5 for awards not planned to occur during CRs versus the 90-day greatest acceleration for awards planned during CRs. And the difference in the lengths of delays between the two cases (219 days versus 189 days) is not statistically significant, which is consistent with the result that the likelihood of experiencing a delay between the two cases is not significantly different.

So, to recap: Between FY 2013 and FY 2015, delays were no more likely for procurement awards planned during periods when CRs subsequently occurred than not, and, when delays occurred in each case, their lengths were not significantly different.

Table 1.6 considers the average change between originally projected unit cost and the actual unit cost for awards projected to occur during and not during CRs between

Table 1.6
Changes in Unit Costs for Procurement Awards Projected to Occur During and Not During Continuing Resolutions for Fiscal Years 2013 Through 2015

Average Change in Unit Cost Relative to Original Projection	Percentage
All services, programs, overall	–3
All services, programs, during	–4
All services, programs, not during	–1

SOURCE: DoD data.

NOTE: Difference between during and not during is not significant ($p = 0.2650$); effect size = 0.1872; 95-percent confidence interval for effect size = [–0.1414, 0.5158].

FY 2013 and FY 2015. Our analysis revealed no statistically significant difference in unit-cost changes between awards planned during or not during CRs. It also showed that, overall, final award costs were, on average, slightly lower than original projections.

Analysis Comparing Awards Planned for a Period Between Fiscal Year 2013 and Fiscal Year 2015 with Awards Planned for a Period in Fiscal Year 1999

To explore whether there might be systematic effects associated with CRs, such as program managers anticipating that CRs would occur and altering their plans on that basis, we performed a limited comparison of changes in projected award dates and unit costs for the 151 procurements planned for some period between FY 2013 and FY 2015 with 48 procurements planned during FY 1999, when only a three-week CR occurred. The results are mixed.

If program managers have been anticipating CRs, one consequence might be that they would plan to make awards later in the year (for example, later than the first quarter of the FY when CRs occur) than they had before long CRs had become common. Between FY 2013 and FY 2015, when long CRs had become common, 93 of the 151 awards we considered, or 62 percent, were planned to occur during a period subsequently affected by a CR. During FY 1999, when long CRs were not common, 25 of 48 awards, or 52 percent, were planned originally to occur during the first quarter of that FY. The differences in these proportions are not statistically significant (Fisher test, $p = 0.3116$, but the number of data points is limited, and significance will be indicated only for relatively large differences). If the majority of all these awards (across all years considered) were being made with prior-year funds, which we cannot discern, program managers would be indifferent to whether CRs were occurring, because those prior-year funds would enable awards to be made irrespective of the occurrence of a CR and irrespective of whether those awards were planned earlier or later in the FY. Whether this is or is not the case, this result—no significant difference in the proportion of awards planned early in the FY between FY 1999 and FYs 2013 through

2015—is at least an indicator that advance planning for CRs by program managers might not be routine.[61]

Of the 48 FY 1999 awards considered in our analysis, 27, or 56 percent, experienced delays. This proportion of delays is not significantly different from the 78 of 151, or 52 percent, of awards spanning FYs 2013 through 2015 that were delayed.[62] However, when delays occurred, they were significantly longer for the awards planned for periods between FY 2013 and FY 2015 than for the awards planned for FY 1999 (see Table 1.7). The latter result could be a systematic effect associated with CRs. However, it could also be associated with changes in acquisition laws, such as WSARA, and regulations that have occurred during the 14 to 16 years intervening between FY 1999 and FY 2013 through FY 2015 for which we have not controlled. Or the result could be due to other factors, including particular features of the programs in our data sets that we do not realize exist and have not considered.

Table 1.7
Comparison of Net Changes in Award Dates and Delays for Procurements Planned for Periods Between Fiscal Year 2013 and Fiscal Year 2015 Versus Those for Periods in Fiscal Year 1999

Metric	Average Change in Award-Days Relative to Original Projection
Net change overall	
All services, FY 2013–FY 2015	95
All services, FY 1999[a]	64
Greatest delay, FY 1999	334
Greatest acceleration, FY 1999	−61
Delays only	
All services, FY 2013–FY 2015	209
All services, FY 1999[b]	124

SOURCE: RAND analysis of data from the Department of Defense.

[a] Difference is not significant ($p = 0.1088$); effect size = 0.2033; 95-percent confidence interval for effect size = [−0.1221, 0.5287].

[b] Difference is significant ($p = 0.0005$); effect size = 0.6193; 95-percent confidence interval for effect size = [0.1737, 1.0650]; this is also reflected in the 334-day greatest delay for FY 1999 awards, versus the 639-day greatest delay for FY 2013–2015 awards (see Table 1.5).

[61] A comprehensive analysis of the steps program managers and acquisition officials might now be routinely taking in anticipation of CRs is beyond the scope of this limited effort

[62] A Fisher test yields $p = 0.6211$, indicating that the two proportions are not significantly different. However, both for this case and for the proportion of awards planned early in the FY, the number of data points is limited, limiting the tests' ability to indicate statistical significance unless differences are relatively large.

Interestingly, Table 1.8 indicates that unit-cost increases relative to original projections appear to be significantly larger for awards planned for some time in FY 1999 than for awards planned for some time between FY 2013 and FY 2015. As discussed previously, this could indicate that the advent of long CRs has provided an incentive for program managers, as well as for industry, to be conservative in estimating unit costs and making bids. But it could also reflect changes in acquisition laws and regulations (such as WSARA) emphasizing realistic cost-estimating, or it could be associated with unique circumstances of particular awards of which we are unaware.

Observations on Specific Awards

The remainder of this section discusses in greater detail selected notable aspects of the data on projected awards we analyzed, focusing on awards that were the subject of draft anomalies prepared by DoD. Although unit-cost increases did occur, particularly for Army programs (in which cost changes ranged from an increase of 83 percent to a decrease of 23 percent), those increases do not appear to be associated with CRs.[63] Similarly, the fact that delays, sometimes substantial, occurred in about one-half the cases we examined is not surprising. Budget materials for the new FY are usually prepared in late fall and submitted to Congress in the early spring of the current FY. Thus, the projections made for executing awards under new or existing contracts in the new FY are estimates prepared 11 to 23 months in advance of when an award would actually be accomplished during the FY for which the budget material is being submitted.[64] When the budget materials are prepared, final decisions to proceed by the responsible acquisition officials in the Office of the Secretary of Defense and the services might not have

Table 1.8
Comparison of Changes in Unit Costs for Procurements Planned for Periods Between Fiscal Year 2013 and Fiscal Year 2015 Versus Those in Fiscal Year 1999

Metric	Average Percentage Change in Unit Cost Relative to Original Projection
All services, FY 2013–FY 2015	–3
All services, FY 1999[a]	19

SOURCE: DoD data.

[a] We consider the difference significant notwithstanding inconsistent indicators—t-test yields $p = 0.1350$; effect size = 0.4372; 95-percent confidence interval for effect size = [0.1096, 0.7648].

[63] In some cases, we reviewed programs' SARs, if produced, for information regarding the programs' histories. The SARs we reviewed provided no information attributing cost increases or any other changes in programs to CRs, although, in one instance (the Apache helicopter remanufacture program), cost increases were attributed in part to congressional action to reduce the President's budget request.

[64] CRs first occur early in a FY, in the fall of the calendar year, and often extend through the Thanksgiving, Christmas, and New Year's holidays. Thus, procurement awards planned for periods subsequently affected by CRs would occur about a year after their original projections are made.

been made, and negotiations with contractors might not have commenced or been completed. Delays in the timing of these decisions relative to early projections, as well as in beginning and concluding contract negotiations, are to be expected, given all of the procedural steps that must be accomplished to satisfy legal and regulatory requirements, as well as the possibility that substantive unforeseen problems (such as negative test results) can emerge as the program proceeds and contract negotiations ensue.

CH-47 Helicopter Multiyear Procurement

The data we analyzed include those for MYP awards and do not indicate that CRs have disrupted such awards. This is noteworthy because anomalies prepared in DoD have expressed the concern that, absent language in the CR permitting specific MYP awards to be made, those multiyear agreements could be broken, and cost increases and production breaks would ensue. For example, an anomaly prepared in anticipation of the FY 2013 CR in DoD for the CH-47 new-build (as opposed to remanufacture) helicopter procurement award under a second, new multiyear contract states that, absent specific permission to proceed with the award in December 2012, the program

> [m]ay incur liability claims for FY12 Advanced Procurement (up to $54.9M) for work in progress, disruption to the production line, and workforce stability; would be forced to revert back to more costly single-year procurements with a loss of $373M in multiyear savings; and a new FY 13 single-year contract award would be delayed 18 to 24 months resulting in a costly gap in production, delaying delivery of aircraft in direct support of worldwide contingency operations.[65]

The FY 2013 President's budget submitted in February 2012 anticipated award of that contract to occur in January 2013 for 25 new-build helicopters. However, the December 2012 SAR for CH-47 states that contract negotiations for the multiyear contract completed December 10, 2012, and the projected contract award date was May 15, 2013. Thus, this appears to be an example in which the early projections in the President's budget of when all the actions, including contract negotiations, would be complete to support award of the contract were optimistic. The timing of the preparation in DoD of the anomaly cited above (which was not included in the FY 2013 CR; see Table 1.2) is not clear, but the date on which DoD anticipated the award to occur was also optimistic. The December 2014 SAR states that the contract was awarded June 10, 2013. Comparing the December 2011, December 2012, and December 2015 SARs indicates that flyaway unit procurement costs for FY 2013 and FY 2014 purchases increased relative to prior estimates by 2 percent to 4 percent.[66] The CR

[65] Office of the Under Secretary of Defense for Acquisition, Technology and Logistics, "Address OSD-ATL Questions on Continuing Resolution Anomalies," document provided to the authors, undated (a).

[66] In the case of CH-47, we use the flyaway unit costs derived from reporting in the December 2011 and December 2012 SARs as the unit costs originally projected for the procurements planned and reported in the FY 2013 and FY 2014 budgets, respectively. The 2011 SAR was prepared at the same time the FY 2013 budget was pre-

appears to have had no effect on the award of this CH-47 multiyear contract, which was awarded within one month of the date projected in the December 2012 SAR and about six months later than projected originally. Although the CR did not affect the award, it did occur later than projections in the FY 2013 budget and the anomaly, and that lateness led to none of the deleterious effects of a delay posited in the anomaly.

The next award under the second CH-47 multiyear contract was anticipated in the FY 2014 President's budget (submitted in April 2013) to occur in May 2014. The award was actually made December 26, 2013.[67] We were not provided with a request prepared in DoD associated with this MYP award, which was the second under the contract first awarded in June 2013. No anomaly addressing this award was included in the FY 2014 CR. The award occurred five months in advance of the originally projected date and was made during the FY 2014 CR. The FY 2013 and FY 2014 CRs appear to have had no effect on the procurement contract awards planned for those two FYs, and the FY 2014 CR did not prevent the second award.

The Global Rapid Response Information Package

The Global Rapid Response Information Package (GRRIP) is a deployable communication package that provides commanders with access to the Secret Internet Protocol Router Network and the Non-Classified Internet Protocol Router Network. The FY 2013 President's budget anticipated an award during that FY for upgrades for 200 GRRIP sets but provided no information regarding the specific award date anticipated or the contract to be initiated or used. An anomaly prepared in DoD—which was not included in the FY 2013 CR—indicated concern that the CR could delay the award, stating that delay would preclude personnel from accessing the "Government network and Information Assurance Vulnerability Assessment [sic] (IAVA) upgrades, [and] creates a security risk for the Command."[68]

The FY 2015 President's budget indicates that 100 GRRIP set upgrade kits were procured in January 2013, during the period of the FY 2013 CR. It provides no information regarding the reason for the reduction in the number of kits procured relative to the number originally anticipated, but the CR did not prevent an award from being made.

---

pared, and the 2012 SAR was prepared at the same time the FY 2014 budget was prepared. We use the flyaway unit costs derived from reporting in the December 2015 SAR as the unit costs actually realized, compared with the procurements planned in both the FY 2013 and FY 2014 budgets. The unit costs reported in the President's budget material for FY 2013 and FY 2014 are more than a factor of two below the SAR costs, as well as more than a factor of two below all the unit costs for the CH-47 new-build program reported for all other years' procurements in the budget material. This discrepancy indicates that the unit costs reported in the FY 2013 and FY 2014 budget material are incorrect.

67 See USAspending.gov, "Advanced Search," undated (b).

68 Office of the Under Secretary of Defense for Acquisition, Technology and Logistics, undated (a).

The AH-64E Apache Attack Helicopter

The FY 2013 President's budget anticipated a procurement award for remanufactured Apache helicopters in October 2012. An anomaly prepared in DoD expressed the concerns that, absent inclusion of the anomaly in the CR, a

> [c]ontract cannot be awarded at the minimum economic production quantity; will cause an increase of the procurement unit price as well as a potential Nunn–McCurdy program breach; will cause a break in the production line disrupting cost and schedule synergies with existing Foreign Military Sales contracts.[69]

During FY 2013, five awards, totaling $127.7 million, were made to the Apache procurement contract the Army used to fund both remanufactured and new-build aircraft, all of which were characterized as associated with advance procurement for Apache full-rate production (FRP).[70] The first of these awards was made on December 21, 2012, during the period of the FY 2013 CR. An award characterized as being for Apache FRP of $585.2 million was made on March 4, 2014, about 17 months after the date projected originally.

The December 2015 SAR for the program displays 37 aircraft procured during FY 2013 and indicates that the only procurement contract in effect with Boeing, the prime contractor for Apache, was the contract to which the awards cited above were made.[71] The SAR also displays an increase in procurement unit costs of about 9 percent, consistent with the FY 2013 and subsequent years' budget material, which indicates about a 10-percent increase in contract unit costs for both remanufactured and new-build Apaches. Regarding costs, the December 2013 SAR states that "fact of life" changes are affecting Apache costs, indicating further that

> [t]hese changes include FY 2013 Congressional reductions ($50M), FY 2013 Sequestration ($34M), an FY 2014 Army adjustment ($100M), and the current FY 2015 PB [President's budget] changes ($175M). These changes significantly reduce

---

69 The Department of Defense Authorization Act, 1982 (Pub L. 97-86, December 1, 1981) includes an amendment sponsored by U.S. Senator Sam Nunn and U.S. Representative Dave McCurdy requiring that major defense acquisition programs be considered for termination and recertified by the Secretary of Defense as essential to national security, if those programs experience an increase in unit costs of 25 percent or more relative to the program's most recent baseline or 50 percent or more relative to the program's original baseline. See Defense Acquisition University, "Nunn–McCurdy Breach," *Defense Acquisition Glossary*, undated.

70 See USAspending.gov, undated (b).

71 The results displayed in Table 1.3 and Table 1.4 show a 17-month delay in contract award for Apache remanufacture based on the information displayed in the President's budget submissions. However, it is clear that fabrication of the 37 FY 2013 helicopters was initiated using the awards made to the procurement contract beginning in December 2012, so the actual practical delay was clearly less than 17 months but cannot be accurately specified based on the information available to us.

> procurement quantities below the minimum sustainment rate of 48 per annum for the AH-64E production line which significantly increases unit costs.[72]

There is no other mention in the 2012 through 2015 SARs of congressional action—including CRs—affecting unit costs or causing production-line breaks or another Nunn–McCurdy breach. (There had been a prior breach as reported in the December 2009 SAR because of the addition to the program of new-build aircraft.) The award of advance procurement funding made during the FY 2013 CR, as well as the SAR reporting, indicates that the concerns expressed in the anomaly were not realized.

Regarding schedules, the December 2012 SAR indicates that the FRP decision for new-build helicopters, which was anticipated in the FY 2013 President's budget to occur in January 2013, was delayed from July 2012 (in the prior SAR) to March 2013. Regarding reviews and approvals within DoD, the SAR notes that the Defense Acquisition Executive (DAE) approved FRP for the remanufacture program but approved only an extended low-rate initial production for the new-build program. The DAE also delegated the Milestone Decision Authority for both the new-build and remanufacture programs to the Army Acquisition Executive (AAE), contingent on the DAE's approval of the remanufacture acquisition program baseline (APB). The DAE approved the remanufacture APB on November 26, 2012. The program manager for Apache, through the program executive officer for aviation, requested an Apache Block IIIB new-build FRP acquisition decision memorandum from the AAE on January 11, 2013. The AAE signed the new-build acquisition decision memorandum for FRP on March 11, 2013, and the APB was in the signature process at that time.

The Joint Air-to-Surface Standoff Missile

Aircraft employ the Joint Air-to-Surface Standoff Missile (JASSM) to attack surface targets at relatively long range. There are baseline and extended-range versions of the missile. The FY 2015 President's budget submitted in March 2014 indicates that award of a production contract was anticipated in December 2014. An anomaly prepared in DoD in anticipation of the FY 2015 CR, but not included in it, expressed the concern that "[n]ot awarding the contract by December 31 [2014] invalidates the negotiated prices; new negotiations will result in increased cost."[73]

A new contract was awarded in October 9, 2015, during the period of the FY 2016 CR, using FY 2014 and FY 2015 funds for 240 JASSMs, up from the originally projected acquisition of 224. Although the award was delayed by almost a year, the President's budget material indicates a decrease in contract unit costs of about 10 percent relative to original projections. The December 2015 SAR states that the produc-

---

[72] U.S. Department of Defense, *AH-64E Apache Remanufacture (AH-64E Remanufacture): December 2015 Selected Acquisition Report*, March 21, 2016, p. 19.

[73] Office of the Under Secretary of Defense for Acquisition, Technology and Logistics, "Appropriation Anomalies," document provided to the authors, undated (b).

tion lot awarded in October for the extended-range version of JASSM resulted in a cost decrease (presumably relative to the immediately prior lot) of 17 percent and that the next lot's costs declined by 22 percent. The next lot was awarded December 1, 2015, under the contract used for the October award, also during the FY 2016 CR.[74] The FY 2014, FY 2015, and FY 2016 CRs had no apparent effect on the program, and the concerns expressed in the anomaly were likely not realized. It should be noted that procurement awards were made under two new separate contracts on December 19, 2013, for the immediately prior (relative to the fall and winter of 2015) lots of both versions of the missile, during the FY 2014 CR.[75]

The MH-60R Helicopter

The Navy uses these Seahawk helicopters to perform antisubmarine warfare, antisurface warfare, and other secondary missions. The FY 2015 President's budget anticipated an award in February 2015 for these helicopters. An anomaly prepared in DoD but not included in the FY 2015 CR indicated the need for the CR to permit an increase in procurement from 19 to 29 helicopters to avoid "[c]ontract delays in FY 2015, impacting production line, delivery schedules, and potentially cost."[76] An award for "program year 4" Navy helicopter procurement was made November 17, 2014, during the FY 2015 CR, to the Army contract under which, according to the December 2015 SAR, these Navy versions of the Sikorsky Blackhawk helicopters have been procured.[77] The SAR indicates that 29 MH-60R helicopters were procured in FY 2015. The President's budget material indicates no increase in contract unit costs relative to original projections. Thus, the concerns raised in the anomaly were not realized.

Rolling Airframe Missile

The Rolling Airframe Missile (RAM) is deployed on the Navy's surface combatants for protection against antiship cruise missile attacks. The FY 2015 President's budget anticipated an award for 90 missiles in December 2014. An anomaly prepared in DoD indicated the need for the CR to permit a production increase from 66 to 90 missiles to avoid delays that would "impact production line, delivery schedules, and potentially cost."[78] The FY 2017 President's budget indicates that a contract award did not occur until February 2015, and deliveries were delayed correspondingly, but the budget indicates that 90 missiles were procured and that contract unit costs decreased by about 20 percent relative to the projections in the FY 2015 President's budget. SARs are not

---

74 See USAspending.gov, undated (b).

75 See USAspending.gov, undated (b).

76 Office of the Under Secretary of Defense for Acquisition, Technology and Logistics, undated (b).

77 See USAspending.gov, undated (b). The FY 2015 budget material indicates incorrectly that the award was made in December 2014.

78 Office of the Under Secretary of Defense for Acquisition, Technology and Logistics, undated (b).

prepared for the RAM. Although we cannot definitively rule out that the CR caused the delay in award from December to February, the unit-cost decrease indicates that the concerns expressed in the anomaly regarding effects on the production line and costs were not realized.

Littoral Combat Ship

Although the budget materials for FY 2013 through 2015 display no delays in awarding littoral combat ship (LCS) construction contracts, the budgets spanning FY 2013 to FY 2018 do display substantial delays in ship deliveries relative to initial projections. For example, the median delay in delivery for LCS hulls 9 through 24 projected in the FY 2018 budget material relative to the projections in the FY 2013 budget is 472 days, ranging from a minimum of 365 days for LCS 18 to a maximum of 578 days for LCS 9. Each successive budget for FYs 2013 through 2018 projects additional delays in delivery for several, if not all, of the 16 ships.

GAO reports have examined these delays; in a June 2016 report, GAO stated,

> LCS cost overruns have been accompanied by significant schedule delays at both shipyards. In 2013, we found that delivery of the two lead ships and LCS 4–8 were delayed by as much as 2 years due to various design and construction issues. At that time, the Navy reported that it had adjusted delivery schedules to account for these delays and did not envision further delays beyond LCS 8. However, our analysis of Navy contracting and budget documents identified that actual or planned deliveries of almost all LCS under contract (LCS 5–26) were delayed by several months, and in some cases close to a year or longer.[79]

GAO also noted that, "[i]n addition to providing schedule relief, through these contract modifications [to the original contracts] the Navy also provided the two shipbuilders with the ability to receive up to $45 million each in incentive fees for launching and delivering LCS."[80]

Regarding cost increases in the LCS program, in the same report, GAO stated,

> The Navy received an additional $160 million in fiscal years 2015 and 2016 and plans to request another $239 million through fiscal year 2020 to complete construction of ships funded in prior years. According to Navy budget documentation, this additional funding is needed to cover the government's share of shipbuilding contract overruns and restoration of de-scoped requirements resulting from sequestration reductions. For fiscal years 2015–2017, of the funding received or requested by the Navy to address LCS program shortfalls ($246 million), almost 70 percent ($169 million) was used or is planned to be used to address shipbuilding

---

[79] GAO, *Littoral Combat Ship: Need to Address Fundamental Weakness in LCS and Frigate Acquisition Strategies*, Washington, D.C., GAO-16-356, June 9, 2016b, p. 35.

[80] GAO, 2016b, p. 35.

> contract overruns on 12 LCS seaframes [LCS 5 through LCS 16; odd-numbered ships are the *Freedom* class and even-numbered ships are the *Independence* class] funded in fiscal years 2010–2013. Additionally, the LCS and frigate programs reported that costs of LCS currently under construction would increase further due to rising shipyard labor rate costs . . . .[81]

Although GAO reports that the Navy does tie LCS cost increases to sequestration, neither GAO's reports nor the program's SARs identify CRs as a causative factor in those increases or delays in LCS deliveries. Previous GAO reports also expressed concern with concurrent design changes and construction of LCSs. For example, in a July 2013 report, GAO stated,

> The Littoral Combat Ship (LCS) seaframe program continues to face challenges stemming from concurrent design, production, and testing activities. The Navy has taken steps to resolve problems with the lead ships, and the shipyards are beginning to realize benefits from facility improvements and experience. However, testing remains to be completed and the Navy is currently studying potentially significant design changes, such as increasing the commonality of systems between the two ship variants and changing ship capabilities. Changes at this point can compromise the positive impacts of shipyard learning, increase costs, and prolong schedules.[82]

An anomaly prepared in DoD anticipating CRs in FY 2016 expressed the concern that, absent special transfer authority enabling funding to be applied to the completion of LCS hulls 9 through 12, "the inability to fund the shortfalls in the first quarter of FY 2016 may slow or stop construction . . . ."[83] However, as discussed, in FY 2015, the Navy had already requested and received funding to pay for contract cost overruns on these ships and anticipated requesting additional funding to cover additional overruns on LCSs. Moreover, issues with slippage in the schedules for delivery of LCS hulls prior to LCS 9 were significant, and, although GAO noted improvements in its July 2013 report, it also cited remaining risks for schedule slippage. It should also be noted that the FY 2015 President's budget submitted in March 2014—the first submitted after the FY 2013 CR—projected the least incremental schedule slippage for LCS hulls 9 through 12 (about 2 percent) of any of the budgets submitted from FY 2013 through FY 2018. If CRs were causing schedule slippage, it would be expected that the

---

81 GAO, 2016b, pp. 33–34.

82 GAO, *Navy Shipbuilding: Significant Investments in the Littoral Combat Ship Continue Amid Substantial Unknowns About Capabilities, Use, and Cost*, Washington, D.C., GAO-13-530, July 22, 2013, "What GAO Found."

83 Office of the Under Secretary of Defense for Acquisition, Technology and Logistics, "FY2016 DoD Appropriation Anomalies," document provided to the authors, undated (c).

President's budget submitted first after one of the longest CRs (178 days) would have projected more-substantial incremental delays in LCS deliveries.

The history of the problems the LCS program has encountered—the continual slippage in delivery schedules projected for LCS hulls 9 through 24 spanning six FYs resulting in delays exceeding one year for each of the 16 ships; the Navy's continued requests for funding to cover past and anticipated future cost overruns; the attribution of cost increases reported in SARs to continued need to fund changes during construction; the lack of GAO or DoD attribution of cost increases and delays in LCS deliveries to CRs; and, in those same GAO and DoD sources, no indication that lack of funding has caused construction to be halted on an LCS—suggests that factors other than CRs have had the most effect on LCS cost increases and delayed deliveries.

Levers for Management Discretion Despite a Continuing Resolution

Managers whose organizations have faced operating under CRs in recent FYs nonetheless do have levers by which to exert a degree of managerial discretion. First, it is important to note that the appropriate application of a CR requires substantial interpretation. The 2006 GAO Red Book on federal appropriation law details many legal interpretations regarding congressional intent on specific projects and activities that have shaped the application of CRs in recent history.[84] The application of a CR is rarely clear cut, and managers might have opportunities to benefit from legal interpretations of a CR's strictures in a more favorable light. Nonetheless, there are limits to legal interpretations, and operating under a CR is likely to impose certain constraints. To address these constraints, managers facing the likelihood of a CR at the beginning of each FY have also reportedly developed techniques for planning ahead and taking proactive steps to mitigate its effects. This section briefly describes some of the levers decisionmakers can employ to advance priorities, even under a CR.

Planning

The extent to which CRs have become the norm for DoD has reportedly shaped the way in which senior leaders manage their organizations. Chief of Naval Operations John M. Richardson testified in a Senate hearing in 2016 that a CR in the first quarter of a new FY is now taken for granted and that, as a result, "[t]he services are essentially operating in three fiscal quarters per year now. Nobody schedules anything important in the first quarter."[85] In short, one technique senior leaders can implement to mitigate the risks introduced by a CR is minimize contract actions planned for the first quarter

---

[84] GAO, 2006.

[85] Williams and Wees, 2016, p. 7.

of a FY.[86] One budget expert, Thad Juszczak at American University, recommended that managers avoid timing recurring annual contract obligations for the first quarter of a fiscal year for precisely this reason.[87] Juszczak also recommended that managers avoid planning new starts for the first quarter and, if that proves to be impossible, have contingency plans in place.[88] Seeking to avoid new starts might lead managers to adjust their organizations' schedules to push such activities as hiring or travel later into the FY. However, this incentive can also drive significant inefficiencies, as, for example, managers risk ending up with a rush to obligate funds (especially one-year money, such as O&M) before the end of the FY.

Managers' ability to put off key decisions until a regular appropriation is in place, however, is a function of the length of the CR. By many accounts, the challenge of managing under a CR grows with the duration of the CR. One budget expert noted that,

> [w]ith proper planning and sound risk management during execution, agencies can operate under a CR with a minimum of disruption, as long as the CR period does not stretch much beyond six weeks. In the case of a longer CR, there will almost certainly be disruptions to business-as-usual.[89]

Unobligated Balances from Prior Years

One lever that can reportedly provide valuable flexibility to managers is the use of unobligated balances from prior years to mitigate the effects of a CR. As CRS analysts explained, "Although appropriations bills most commonly provide budget authority that is available for obligation for only one fiscal year, budget authority for an activity may be provided for more than one year ('multiyear') or indefinitely ('no-year')."[90] These multiyear funds can result in balances available for obligation in subsequent years without additional appropriation. Juszczak explained that a wise manager will conserve such unobligated funds because their availability "for emergencies like a CR in the new fiscal year can provide real flexibility for an agency."[91] Similarly, a CRS report noted that, despite the prohibition on new starts under a CR, "[a]gencies may use appropriated funds from prior fiscal years that remain available . . . to initiate new

86 Williams and Wees, 2016, p. 7.

87 Thad Juszczak, "Continuing Resolutions Don't Have to Be a Pain," *govloop*, October 14, 2009.

88 Juszczak, 2009.

89 Juszczak, 2009.

90 Saturno and Tollestrup, 2016, p. 7, footnote 21.

91 Juszczak, 2009.

activities in some circumstances."[92] Notably, however, not all programs have this flexibility. For example, programs driven by one-year appropriations, such as O&M, would generally need to be obligated in the year appropriated.

Reprogramming

Another lever the Pentagon maintains for exercising a degree of managerial discretion is reprogramming of previously received appropriations toward new purposes. Reprogramming is defined as the "use of funds for a project or purpose other than that for which they were originally provided."[93] In 2017, the Deputy Assistant Secretary of the Navy for RDT&E, William P. Bray, reported that

> reprogramming funds provides one way for the Marine Corps to prioritize programs based on timing . . . . For example, if I've got program A over here and it's an important thing, but I've got something I can get out there quicker, if I take money from here and move it, it doesn't mean that this program is less important than what it was. It means you've got something here that I can field quickly . . . . In a zero-sum game, you've got to make those decisions, and the service has to have the flexibility to go do that.[94]

However, a decision to reprogram funds needs to reflect congressional intent. Appropriators in Congress hold the power of the purse, so a department request to reprogram funds requires engagement with Capitol Hill. Most appropriation committees require notification of intent to reprogram, and most also require congressional approval for reprogramming above a certain threshold.[95] The Pentagon, for example, "requires approval for reprogramming above $20 million, or 2 percent of the appropriated amount, whichever is greater, according to one budget expert."[96] In short, the process is time-intensive and the outcome uncertain.

---

[92] Saturno and Tollestrup, 2016, p. 7. Moreover, our data include cases in which awards made during CRs were, not surprisingly, made with prior-year funds. However, data on the FY in which the appropriations used to make the awards we examined are not consistently available. Therefore, we cannot comprehensively analyze the extent to which the awards we examined did or did not use prior-year funds.

[93] Michelle D. Christensen, *The Executive Budget Process: An Overview*, Washington, D.C.: Congressional Research Service, R42633, July 27, 2012, p. 8.

[94] "Critical Intelligence," *Inside the Pentagon*, Vol. 33, No. 18, May 4, 2017.

[95] Michelle Mrdeza and Kenneth Gold, *Reprogramming Funds: Understanding the Appropriators' Perspective*, Government Affairs Institute at Georgetown University, undated.

[96] Mrdeza and Gold, undated.

Conclusion

There is general agreement among those who have studied them that CRs, because they limit government agencies' ability to engage in activities for which authority has not been granted by law previously, can cause inefficiency in government operations and have other anticipated harmful effects. Assertions are also commonly made that these anticipated harmful effects can include, among other things, costs for work on the part of government personnel that would not otherwise be done, delays in changing the course of programs or eliminating outmoded programs, and costs for delayed procurements. However, as GAO has noted, data and other information are lacking that would permit analysts to be able to derive defensible estimates of those costs and to draw general conclusions.

In this effort, we focused on just two of the many potential deleterious effects of CRs discussed in past, generally testimony-based, assessments: delays and increases in the costs of planned weapon procurements. We chose this focus for our analysis because data were available to perform it: the budget material displaying projected procurement contract award dates and unit costs composing DoD's President's budget submissions. The concern that CRs would cause delays in making weapon procurement awards under new or existing contracts that would, in turn, increase costs is expressed in some anomalies prepared in DoD during the past several years. Generally, the permissions sought in these anomalies to take specific procurement actions have not been included in the CRs that have been enacted.

In our statistical analysis, which contains mixed results, we did not find strong evidence in the budget materials or SARs indicating that CRs are generally associated with delays in procurement awards or increased costs. On the other hand, given the limitations inherent in our statistical analysis, we cannot use its results to rule out the occurrence of these kinds of negative effects.

In the instances we examined in which DoD staff explicitly expressed concerns in draft anomalies that CRs would have negative consequences, the budget materials and SARs provide evidence that, when negative programmatic effects occurred, they were most likely associated with factors other than CRs. We found that sufficient flexibility exists under certain circumstances for awards under existing multiyear contracts to be made during CRs (e.g., CH-47), as well as for other awards under new contracts to be made, at least for programs that have already entered production and procurement (e.g., JASSM).

This limited effort does not constitute a legal analysis. In particular, in this report, we make no attempt to identify or speculate on the legal rationale that has enabled DoD to make the procurement awards during CRs that have occurred. (For example, use of funds appropriated in prior years could enable an award to be made during a CR, but the data available do not always specify the FY in which the funds awarded were appropriated.) However, DoD acquisition staffs have prepared anomalies based on concerns that some awards, such as those that have been made, could not be made. Therefore, a legal analysis explaining the rationale for the awards that have been made during CRs and distinguishing them from other awards that were (or would have been) delayed could be useful and informative.

Another caveat is that our limited research effort has not examined all the procurement awards that DoD anticipated making during FYs 2013 through 2015 or in prior or subsequent years. We also have not examined whether the concerns expressed in all of the planned procurement anomalies that DoD drafted in anticipation of CRs during the past several years have been realized. Finally, we have performed only a limited-scope analysis of potential systemic effects CRs might be causing. Thus, we cannot conclusively rule out that CRs have had substantial effects on some planned procurement awards or that CRs are having some systemic effects.

For these reasons and others, the lack of conclusive evidence we have found in this particular case does not imply that the widely expressed concerns regarding CR effects are invalid. As the examples described demonstrate, many factors affect how defense programs proceed and how their yearly procurement plans unfold relative to advance projections. Priorities for resource allocation continually evolve within DoD and Congress, driven by many factors, including problems that arise during production of weapons occurring concurrently with completion of design and testing (e.g., LCS and JSF), reviews and certifications within DoD that take longer to complete than hoped (e.g., Apache), and contract negotiations that take longer to resolve than anticipated (e.g., Apache and JSF).[97] Thus, it would be incorrect to reach general conclusions about the effects of CRs on DoD-wide or government-wide operations based only on DoD's experience in the one particular regard we have examined.

---

[97] The problems that the JSF program has encountered as a result of concurrency are documented in numerous sources. For example, GAO has stated, "In our prior work, we identified the lack of knowledge and high levels of concurrency as major drivers in the significant cost and schedule growth as well as performance shortfalls that the program has experienced since 2001" (GAO, *F-35 Joint Strike Fighter: Continued Oversight Needed as Program Plans to Begin Development of New Capabilities*, Washington, D.C., GAO-16-390, April 14, 2016a, p. 3).

Negotiations to award JSF procurement contracts have frequently been delayed relative to original projections and have been contentious. For example, in November 2016, DoD issued a unilateral contract action for production of the lot 9 F-35s after it failed to reach agreement with the prime contractor on price. See Mike Stone, "Pentagon, Lockheed Finalize Talks on Ninth Lot of F-35 Jets," Reuters, November 2, 2016.

APPENDIX

Program Procurement Awards Included in This Assessment

This appendix lists, by service, the awards we included in our assessment. We considered a total of 199 programs planned for FY 1999 and FYs 2013 through 2015.

Air Force

- AC-130 Recapitalization
- Advanced Medium Range Air-to-Air Missile
- AIM-9X Sidewinder
- C-17
- CV-22 Osprey
- F-22
- Global Positioning System III Satellite (satellites 03–08)
- HC-130 Recapitalization
- JASSM
- JASSM–Extended Range
- Joint Direct Attack Munition
- Joint Primary Aircraft Training System
- Joint Standoff Weapon
- JSF
- Mark 84 Bomb
- MC-130 Recapitalization
- MQ-9 Reaper
- Sensor Fuzed Weapon
- Small Diameter Bomb Increment II
- Wind Corrected Munitions Dispenser

Army

- (CH-47) Chinook Multiyear New-Build
- Abrams Tank
- AH-64 Apache Block IIIA Remanufacture
- AH-64 Apache Block IIIB New-Build
- AN/ARC-220 Radio
- AN/PSC-15 GRRIP
- AN/VR-100 Radio
- Army Tactical Missile System
- Assault Breacher Vehicle
- Avenger
- Bradley Fighting Vehicle
- Command and Control Vehicle
- Counterfire Radars
- Fire Support Team Vehicle
- General Fund Enterprise Business System–Sensitive Activities
- Guided Multiple Launch Rocket System
- Heavy Assault Bridge
- Improved Recovery Vehicle (M88A2 HERCULES [Heavy Equipment Recovery Combat Utility Lift and Evacuation System])
- Javelin
- Light Utility Helicopter
- Lightweight Counter Mortar Radar
- Longbow Hellfire
- MQ-1 Unmanned Aerial Vehicle
- Paladin Integrated Management
- Patriot Missile
- Prophet Ground
- Reserve Component Automation System
- RQ-11 Raven
- Secure, Mobile, Anti-Jam, Reliable, Tactical–Terminal
- Signal Modernization
- Stryker Vehicle
- Tactical Airspace Integration System
- Tube-Launched, Optically-Tracked, Wire-Guided Missile
- UH-60M Blackhawk
- Warfighter Information Network–Tactical, Increment 1

Navy and Marine Corps

- Advanced Medium-Range Air-to-Air Missile
- AIM-9X Sidewinder Air-to-Air Missile
- Airborne Laser Mine Detection System
- AV-8B
- CH-60 Helicopter
- DDG 1002 Multi-Function Radar
- DDG 51 Class
- Distributed Common Ground System—Marine Corps Selected Equipment
- E-2C Hawkeye Aircraft
- E-2D Advanced Hawkeye Aircraft
- EA-18G Growler
- Evolved Sea Sparrow Missile
- F/A-18 E/F Super Hornet
- Joint High Speed Vessel
- Joint Primary Aircraft Training System
- Joint Standoff Weapon
- Landing Craft, Air-Cushioned Service Life Extension
- LCS *Freedom* Class
- LCS *Independence* Class
- MH-60R Helicopter
- MH-60S Helicopter
- MK 48 Common Broadband Advanced Sonar Heavyweight Torpedo Kits
- P-8A Poseidon
- Pioneer Unmanned Aerial Vehicle
- RAM
- RQ-11 Unmanned Aerial Vehicle
- RQ-21 Unmanned Aerial System
- Standard Missile 2 Block IV
- Standard Missile 6
- T-45 Trainer Aircraft
- Tactical Tomahawk for Vertical Launching System
- Total Ship Computing Environment for DDG 1002
- Trident II D5 Solid Rocket Motors
- Trident II D5 Strategic Program Alteration Kits
- UH-1Y Helicopter
- V-22 Osprey
- *Virginia* Class Submarine Command, Control, Communications, and Intelligence System Shipsets

Bibliography

"Acquisition Process: Selected Acquisition Report (SAR)," *AcqNotes*, updated July 11, 2017. As of April 30, 2018:
http://acqnotes.com/acqnote/acquisitions/selected-acquisition-report-sar

"Appropriations for Fiscal Year 2017," *Congress.gov*, last updated May 4, 2017. As of May 12, 2017:
https://www.congress.gov/resources/display/content/Appropriations+for+Fiscal+Year+2017

Belasco, Amy, and Pat Towell, *Government Shutdown: Operations of the Department of Defense During a Lapse in Appropriations*, Washington, D.C.: Congressional Research Service, R41745, October 15, 2013. As of April 30, 2018:
https://fas.org/sgp/crs/natsec/R41745.pdf

Bond, Craig A., Lauren A. Mayer, Michael McMahon, James G. Kallimani, and Ricardo Sanchez, *Developing a Methodology for Risk-Informed Trade-Space Analysis in Acquisition*, Santa Monica, Calif.: RAND Corporation, RR-701-A, 2015. As of June 6, 2018:
https://www.rand.org/pubs/research_reports/RR701.html

Brass, Clinton T., *Interim Continuing Resolutions (CRs): Potential Impacts on Agency Operations*, Washington, D.C.: Congressional Research Service, RL34700, March 16, 2010. As of April 30, 2018:
http://www.dtic.mil/get-tr-doc/pdf?AD=ADA517371

Brennan, Rick, Jr., Charles P. Ries, Larry Hanauer, Ben Connable, Terrence Kelly, Michael McNerney, Stephanie Young, Jason H. Campbell, and K. Scott McMahon, *Ending the U.S. War in Iraq: The Final Transition, Operational Maneuver, and Disestablishment of United States Forces–Iraq*, Santa Monica, Calif.: RAND Corporation, RR-232-USFI, 2013. As of April 30, 2018:
https://www.rand.org/pubs/research_reports/RR232.html

Censer, Marjorie, "Contractors Remain Upbeat on Defense Spending—but Warn of Uncertainty," *Inside the Navy*, Vol. 30, No. 17, May 1, 2017.

Christensen, Michelle D., *The Executive Budget Process: An Overview*, Washington, D.C.: Congressional Research Service, R42633, July 27, 2012. As of April 30, 2018:
https://fas.org/sgp/crs/misc/R42633.pdf

Committee on Armed Services, U.S. House of Representatives, *Damage to the Military from a Continuing Resolution: Hearing Before the Committee on Armed Services*, hearing, Washington, D.C., April 5, 2017. As of April 30, 2018:
https://armedservices.house.gov/legislation/hearings/damage-military-continuing-resolution

"Critical Intelligence," *Inside the Pentagon*, Vol. 33, No. 18, May 4, 2017.

DeBruyne, Nese F., *Defense Authorization and Appropriations Bills: FY1970–FY2017*, Washington, D.C.: Congressional Research Service, January 18, 2017.

Defense Acquisition University, "Nunn–McCurdy Breach," *Defense Acquisition Glossary*, undated. As of May 1, 2018:
https://www.dau.mil/glossary/Pages/2306.aspx

Federal Acquisition Regulation, Part 43, Contract Modifications, Subpart 43.2, Change Orders, last updated January 24, 2018. As of May 1, 2018:
https://www.acquisition.gov/browsefar#

GAO—*See* U.S. Government Accountability Office.

Goldfein, Gen. David L., Chief of Staff, U.S. Air Force, "Department of the Air Force Presentation to the Committee on Armed Services, United States Congress; Subject: Impacts of a Year-Long Continuing Resolution; Statement of: General David L. Goldfein, Chief of Staff of the Air Force," Washington, D.C.: U.S. House of Representatives, April 5, 2017. As of April 30, 2018:
https://docs.house.gov/meetings/AS/AS00/20170405/105832/

HHRG-115-AS00-Wstate-GoldfeinD-20170405.pdf

Hawkings, David, "What Is a Continuing Resolution?" *Roll Call*, September 7, 2016. As of April 30, 2018:
https://www.rollcall.com/policy/what-is-a-continuing-resolution

Joyce, Philip G., *The Costs of Budget Uncertainty: Analyzing the Impact of Late Appropriations*, Washington, D.C.: IBM Center for the Business of Government, Improving Performance series, 2012. As of April 30, 2018:
http://www.businessofgovernment.org/report/
costs-budget-uncertainty-analyzing-impact-late-appropriations

Judson, Jen, "Fanning: Continuing Resolution Could Snarl European Reassurance Initiative Efforts," *DefenseNews*, October 2, 2016. As of April 30, 2018:
https://www.defensenews.com/digital-show-dailies/ausa/2016/10/03/
fanning-continuing-resolution-could-snarl-european-reassurance-initiative-efforts/

Juszczak, Thad, "Continuing Resolutions Don't Have to Be a Pain," *govloop*, October 14, 2009. As of April 30, 2018:
https://www.govloop.com/community/blog/continuing-resolutions-dont-have-to-be-a-pain/

Karas, Rachel S., "Long-Term CR Could Delay Compass Call Fielding by at Least a Year," *Inside the Air Force*, Vol. 28, No. 17, April 28, 2017.

Light, Thomas, Robert S. Leonard, Julia Pollak, Meagan L. Smith, and Akilah Wallace, *Quantifying Cost and Schedule Uncertainty for Major Defense Acquisition Programs (MDAPs)*, Santa Monica, Calif.: RAND Corporation, RR-1723-AF, 2017. As of June 6, 2018:
https://www.rand.org/pubs/research_reports/RR1723.html

Lorell, Mark A., Robert S. Leonard, and Abby Doll, *Extreme Cost Growth: Themes from Six U.S. Air Force Major Defense Acquisition Programs*, Santa Monica, Calif.: RAND Corporation, RR-630-AF, 2015. As of June 6, 2018:
https://www.rand.org/pubs/research_reports/RR630.html

Lorell, Mark A., Leslie Adrienne Payne, and Karishma R. Mehta, *Program Characteristics That Contribute to Cost Growth: A Comparison of Air Force Major Defense Acquisition Programs*, Santa Monica, Calif.: RAND Corporation, RR-1761-AF, 2017. As of June 6, 2018:
https://www.rand.org/pubs/research_reports/RR1761.html

Martin, Belva M., acting director, acquisition and sourcing management, U.S. Government Accountability Office, "Navy's Proposed Dual Award Acquisition Strategy for Littoral Combat Ship Program," memorandum to congressional committees, Washington, D.C., GAO-11-249R, December 8, 2010. As of May 1, 2018:
https://www.gao.gov/products/GAO-11-249R

Milley, GEN Mark A., Chief of Staff, U.S. Army, "Record Version: Statement by General Mark A. Milley, Chief of Staff, United States Army, Before the Committee on Armed Services, U.S. House of Representatives, First Session, 115th Congress, on Consequences to the Military of a Continuing Resolution," Washington, D.C.: U.S. House of Representatives, April 5, 2017. As of April 30, 2018:
https://docs.house.gov/meetings/AS/AS00/20170405/105832/HHRG-115-AS00-Wstate-MilleyM-20170405.pdf

Mrdeza, Michelle, and Kenneth Gold, "Reprogramming Funds: Understanding the Appropriators' Perspective," Washington, D.C.: Government Affairs Institute at Georgetown University, undated. As of April 30, 2018:
https://gai.georgetown.edu/reprogramming-funds-understanding/

Neller, Gen Robert B., Commandant, U.S. Marine Corps, "Statement of General Robert B. Neller, Commandant of the Marine Corps, Before the House Armed Services Committee on Continuing Resolution Consequences," Washington, D.C.: U.S. House of Representatives, April 5, 2017. As of April 30, 2018:
https://docs.house.gov/meetings/AS/AS00/20170405/105832/HHRG-115-AS00-Wstate-NellerR-20170405.pdf

Office of the Secretary of Defense and Joint Staff, "Office of the Secretary of Defense and Joint Staff Freedom of Information Act Requester Service Center," undated. As of April 30, 2018:
http://www.esd.whs.mil/FOID/foi/Reading_Room/Selected_Acquisition_Reports/

Office of the Under Secretary of Defense for Acquisition, Technology and Logistics, "Address OSD-ATL Questions on Continuing Resolution Anomalies," document provided to the authors, undated (a).

Office of the Under Secretary of Defense for Acquisition, Technology and Logistics, "Appropriation Anomalies," document provided to the authors, undated (b).

Office of the Under Secretary of Defense for Acquisition, Technology and Logistics, "FY2016 DoD Appropriation Anomalies," document provided to the authors, undated (c).

Office of the Under Secretary of Defense (Comptroller), "DoD Budget Request," undated. As of April 30, 2018:
http://comptroller.defense.gov/Budget-Materials/

O'Rourke, Ronald, and Moshe Schwartz, *Multiyear Procurement (MYP) and Block Buy Contracting in Defense Acquisition: Background and Issues for Congress*, Washington, D.C.: Congressional Research Service, R41909, March 23, 2016. As of April 30, 2018:
http://www.dtic.mil/get-tr-doc/pdf?AD=AD1007030

Public Law 97-86, Department of Defense Authorization Act, 1982, December 1, 1981.

Public Law 111-23, Weapon Systems Acquisition Reform Act of 2009, May 22, 2009. As of April 30, 2018:
https://www.gpo.gov/fdsys/pkg/PLAW-111publ23/content-detail.html

Public Law 111-242, Continuing Appropriations Act, 2011, September 30, 2010. As of April 30, 2018:
https://www.gpo.gov/fdsys/pkg/PLAW-111publ242/content-detail.html

Public Law 111-290, Joint Resolution Making Further Continuing Appropriations for Fiscal Year 2011, and for Other Purposes, December 4, 2010. As of April 30, 2018:
https://www.gpo.gov/fdsys/pkg/PLAW-111publ290/content-detail.html

Public Law 111-317, Joint Resolution Making Further Continuing Appropriations for Fiscal Year 2011, and for Other Purposes, December 18, 2010. As of April 30, 2018:
https://www.gpo.gov/fdsys/pkg/PLAW-111publ317/content-detail.html

Public Law 111-322, Continuing Appropriations and Surface Transportation Extensions Act, 2011, December 22, 2010. As of April 30, 2018:
https://www.gpo.gov/fdsys/pkg/PLAW-111publ322/content-detail.html

Public Law 112-4, Furthering Continuing Appropriations Amendments, 2011, March 2, 2011. As of April 30, 2018:
https://www.gpo.gov/fdsys/pkg/PLAW-112publ4/content-detail.html

Public Law 112-6, Additional Continuing Appropriations Amendments, 2011, March 18, 2011. As of April 30, 2018:
https://www.gpo.gov/fdsys/pkg/PLAW-112publ6/content-detail.html

Public Law 112-8, Further Additional Continuing Appropriations Amendments, 2011, April 9, 2011. As of April 30, 2018:
https://www.gpo.gov/fdsys/pkg/PLAW-112publ8/content-detail.html

Public Law 112-10, Department of Defense and Full-Year Continuing Appropriations Act, 2011, April 15, 2011. As of June 28, 2018:
https://www.gpo.gov/fdsys/pkg/PLAW-112publ10/content-detail.html

Public Law 112-33, Continuing Appropriations Act, 2012, September 30, 2011. As of April 30, 2018:
https://www.gpo.gov/fdsys/pkg/PLAW-112publ33/content-detail.html

Public Law 112-36, Continuing Appropriations Act, 2012, October 5, 2011. As of April 30, 2018:
https://www.gpo.gov/fdsys/pkg/PLAW-112publ36/content-detail.html

Public Law 112-55, Consolidated and Further Continuing Appropriations Act, 2012, November 18, 2011. As of April 30, 2018:
https://www.gpo.gov/fdsys/pkg/PLAW-112publ55/content-detail.html

Public Law 112-67, Joint Resolution Making Further Continuing Appropriations for Fiscal Year 2012, and for Other Purposes, December 16, 2011. As of April 30, 2018:
https://www.gpo.gov/fdsys/pkg/PLAW-112publ67/content-detail.html

Public Law 112-68, Joint Resolution Making Further Continuing Appropriations for Fiscal Year 2012, and for Other Purposes, December 17, 2011. As of April 30, 2018:
https://www.gpo.gov/fdsys/pkg/PLAW-112publ68/content-detail.html

Public Law 112-74, Consolidated Appropriations Act, 2012, December 23, 2011. As of April 30, 2018:
https://www.gpo.gov/fdsys/pkg/PLAW-112publ74/content-detail.html

Public Law 112-175, Continuing Appropriations Resolution, 2013, September 28, 2012. As of April 30, 2018:
https://www.gpo.gov/fdsys/pkg/PLAW-112publ175/content-detail.html

Public Law 113-6, Consolidated and Further Continuing Appropriations Act, 2013, March 26, 2013. As of April 30, 2018:
https://www.gpo.gov/fdsys/pkg/PLAW-113publ6/content-detail.html

Public Law 113-39, Pay Our Military Act, September 30, 2013. As of April 30, 2018:
https://www.gpo.gov/fdsys/pkg/PLAW-113publ39/content-detail.html

Public Law 113-44, Department of Defense Survivor Benefits Continuing Appropriations Resolution, 2014, October 10, 2013. As of April 30, 2018:
https://www.gpo.gov/fdsys/pkg/PLAW-113publ44/content-detail.html

Public Law 113-46, Continuing Appropriations Act, 2014, October 17, 2013. As of April 30, 2018:
https://www.gpo.gov/fdsys/pkg/PLAW-113publ46/content-detail.html

Public Law 113-73, Joint Resolution Making Further Continuing Appropriations for Fiscal Year 2014, and for Other Purposes, January 15, 2014. As of April 30, 2018:
https://www.gpo.gov/fdsys/pkg/PLAW-113publ73/content-detail.html

Public Law 113-164, Continuing Appropriations Resolution, 2015, September 19, 2014. As of April 30, 2018:
https://www.gpo.gov/fdsys/pkg/PLAW-113publ164/content-detail.html

Public Law 113-202, Joint Resolution Making Further Continuing Appropriations for Fiscal Year 2015, and for Other Purposes, December 12, 2014. As of April 30, 2018:
https://www.gpo.gov/fdsys/pkg/PLAW-113publ202/content-detail.html

Public Law 113-203, Joint Resolution Making Further Continuing Appropriations for Fiscal Year 2015, and for Other Purposes, December 13, 2014. As of April 30, 2018:
https://www.gpo.gov/fdsys/pkg/PLAW-113publ203/content-detail.html

Public Law 114-53, TSA Office of Inspection Accountability Act of 2015, September 30, 2015. As of April 30, 2018:
https://www.gpo.gov/fdsys/pkg/PLAW-114publ53/content-detail.html

Public Law 114-96, Further Continuing Appropriations Act, 2016, December 11, 2015. As of April 30, 2018:
https://www.gpo.gov/fdsys/pkg/PLAW-114publ96/content-detail.html

Public Law 114-100, Joint Resolution Making Further Continuing Appropriations for Fiscal Year 2016, and for Other Purposes, December 16, 2015. As of April 30, 2018:
https://www.gpo.gov/fdsys/pkg/PLAW-114publ100/content-detail.html

Public Law 114-223, Continuing Appropriations and Military Construction, Veterans Affairs, and Related Agencies Appropriations Act, 2017, and Zika Response and Preparedness Act, September 29, 2016. As of April 30, 2018:
https://www.gpo.gov/fdsys/pkg/PLAW-114publ223/content-detail.html

Public Law 114-254, An Act Making Appropriations for Energy and Water Development and Related Agencies for the Fiscal Year Ending September 30, 2016, and for Other Purposes, December 10, 2016. As of April 30, 2018:
https://www.gpo.gov/fdsys/pkg/PLAW-114publ254/content-detail.html

Public Law 115-30, Joint Resolution Making Further Continuing Appropriations for Fiscal Year 2017, and for Other Purposes, April 28, 2017. As of April 30, 2018:
https://www.gpo.gov/fdsys/pkg/PLAW-115publ30/content-detail.html

Sager, Michelle, director, strategic issues, U.S. Government Accountability Office, *Budget Issues: Effects of Budget Uncertainty from Continuing Resolutions on Agency Operations—Statement for the Record by Michelle Sager, Director, Strategic Issues*, Washington, D.C.: U.S. Government Accountability Office, GAO-13-464T, March 13, 2013. As of April 30, 2018:
https://www.gao.gov/products/GAO-13-464T

Samuelsohn, Darren, "Welcome to CR Hell," *Politico*, October 21, 2015. As of April 30, 2018:
https://www.politico.com/agenda/story/2015/10/congress-federal-budget-continuing-resolution-000270

Saturno, James V., and Jessica Tollestrup, *Continuing Resolutions: Overview of Components and Recent Practices*, Washington, D.C.: Congressional Research Service, R42647, January 14, 2016. As of April 30, 2018:
https://fas.org/sgp/crs/misc/R42647.pdf

Shea, Dana A., and Daniel Morgan, *Federal Research and Development Funding: Possible Impacts of Operating Under a Continuing Resolution*, Washington, D.C.: Congressional Research Service, RS22978, October 27, 2008. As of April 30, 2018:
https://digital.library.unt.edu/ark:/67531/metacrs10815/

Stone, Mike, "Pentagon, Lockheed Finalize Talks on Ninth Lot of F-35 Jets," Reuters, November 2, 2016. As of May 1, 2018:
https://www.reuters.com/article/us-lockheed-pentagon-contract-f35-exclus-idUSKBN12X2LK?il=0

USAspending.gov, home page, undated (a). As of April 30, 2018:
https://www.usaspending.gov/

USAspending.gov, "Advanced Search," undated (b). As of May 1, 2018:
https://www.usaspending.gov/#/search

U.S. Code, Title 10, Armed Forces, Subtitle A, General Military Law, Part IV, Service, Supply, and Procurement, Chapter 144, Major Defense Acquisition Programs, Section 2430, Major Defense Acquisition Program Defined. As of May 1, 2018:
https://www.gpo.gov/fdsys/granule/USCODE-2006-title10/USCODE-2006-title10-subtitleA-partIV-chap144-sec2430

U.S. Department of Defense, *AH-64E Apache Remanufacture (AH-64E Remanufacture): December 2015 Selected Acquisition Report*, March 21, 2016. As of June 6, 2018:
http://www.esd.whs.mil/Portals/54/Documents/FOID/Reading%20Room/Selected_Acquisition_Reports/16-F-0402_DOC_27_AH-64E_Remanufacture_DEC_2015_SAR.pdf

U.S. Government Accountability Office, *Principles of Federal Appropriations Law*, 3rd ed., Vol. II, Washington, D.C., GAO-06-382SP, February 1, 2006. As of April 30, 2018:
https://www.gao.gov/products/GAO-06-382SP

U.S. Government Accountability Office, *Defense Acquisitions: DoD's Practices and Processes for Multiyear Procurement Should Be Improved*, Washington, D.C., GAO-08-298, February 7, 2008. As of April 30, 2018:
https://www.gao.gov/products/GAO-08-298

U.S. Government Accountability Office, *Continuing Resolutions: Uncertainty Limited Management Options and Increased Workload in Selected Agencies—Report to the Ranking Member, Subcommittee on Oversight of Government Management, the Federal Workforce, and the District of Columbia, Committee on Homeland Security and Governmental Affairs, U.S. Senate*, Washington, D.C., GAO-09-879, September 2009. As of April 30, 2018:
https://www.gao.gov/assets/300/295970.pdf

U.S. Government Accountability Office, *Navy Shipbuilding: Significant Investments in the Littoral Combat Ship Continue Amid Substantial Unknowns About Capabilities, Use, and Cost*, Washington, D.C., GAO-13-530, July 22, 2013. As of May 1, 2018:
https://www.gao.gov/products/GAO-13-530

U.S. Government Accountability Office, *F-35 Joint Strike Fighter: Continued Oversight Needed as Program Plans to Begin Development of New Capabilities*, Washington, D.C., GAO-16-390, April 14, 2016a. As of May 1, 2018:
https://www.gao.gov/products/GAO-16-390

U.S. Government Accountability Office, *Littoral Combat Ship: Need to Address Fundamental Weakness in LCS and Frigate Acquisition Strategies*, Washington, D.C., GAO-16-356, June 9, 2016b. As of May 1, 2018:
https://www.gao.gov/products/GAO-16-356

Williams, Lynn M., and Darren P. Wees, *FY2017 Defense Spending Under an Interim Continuing Resolution (CR): In Brief*, Washington, D.C.: Congressional Research Service, R44636, November 7, 2016. As of April 30, 2018:
http://www.dtic.mil/get-tr-doc/pdf?AD=AD1021221

Younossi, Obaid, Mark V. Arena, Kevin Brancato, John Graser, Benjamin W. Goldsmith, Mark A. Lorell, Fred Timson, and Jerry Sollinger, *F-22A Multiyear Procurement Program: An Assessment of Cost Savings*, Santa Monica, Calif.: RAND Corporation, MG-664-OSD, 2007. As of April 30, 2018:
https://www.rand.org/pubs/monographs/MG664.html